Cases
in
Systems Design

West Series in Data Processing and Information Systems

V. Thomas Dock, *Consulting Editor*

Cases
in
Systems Design

JAMES C. WETHERBE
University of Houston

West Publishing Company
St. Paul • New York • Los Angeles • San Francisco

Copyright © 1979 by WEST PUBLISHING CO.
50 West Kellogg Boulevard
P.O. Box 3526
St. Paul, Minnesota 55165

Printed in the United States of America

Library of Congress Cataloging in Publication Data

Wetherbe, James C
  Cases in systems design.

  (The West series in data processing and
information systems)
  1. Management information systems—Problems,
exercises, etc. 2. System analysis—Problems,
exercises, etc. I. Title. II. Series.
T58.6.W46   658.4'032   79-1459

ISBN 0-8299-0229-5

TO SMOKY

Contents

Preface

A particularly perplexing problem when developing a meaningful learning experience in systems analysis and design is the difficulty of providing an industry or applied orientation. Systems analysis is an applied discipline. The learning experience is compromised when theories and concepts are only discussed, and are not applied to industry-oriented problems.

Mission of the Book

This book was written to accomplish two missions. First, it may be used as a case book in conjunction with another text, SYSTEMS ANALYSIS FOR COMPUTER-BASED INFORMATION SYSTEMS, West Publishing, written by this author. Second, it may be used as a stand-alone book, to provide specific, efficient, yet comprehensive training in designing information systems specifications. When used in conjunction with SYSTEMS ANALYSIS FOR COMPUTER-BASED INFORMATION SYSTEMS, this case book can be introduced after completing Chapter 6 of the SYSTEMS ANALYSIS text. Chapter 6 introduces the concepts and application of a general-purpose framework for designing systems specifications. Students can select and work on one or more cases from this book while they complete the remainder of the SYSTEMS ANALYSIS text.

Chapter 6 of SYSTEMS ANALYSIS FOR COMPUTER-BASED INFORMATION SYSTEMS is included (with some modifications) as Chapter 1 of this book. This allows the case book to be used as an independent text. The author and several colleagues have found this independent characteristic to be most helpful. The systems specifications framework and the cases can be used in other courses where students can benefit from a pragmatic exposure to systems design basics. In particular, the case book is useful in graduate Management Information System courses. These courses tend to be dominated by concepts and theories and tend to be light on practical skills. Students who do one or more of the cases in this book acquire a more substantive grasp of the specifics required to articulate information systems specifications.

Developing Industry-Based Cases

The motive for developing industry-based systems design cases for students of systems analysis is not dissimilar to the motive for developing "work sets" for accounting students. The primary objective is to provide a comprehensive environment where students can apply their newly learned skills. A problem with developing such cases is that they must be operational within the time constraints of a one- or two-semester course in systems analysis.

The author has developed industry-based cases in a manner analogous to the method used for developing "connect-the-dots" drawings. In "connect-the-dots" drawings, a drawing is first completed; dots are strategically located on the lines of the drawing; and then the lines are removed, leaving only the dots (which can be reconnected to complete the drawing).

To develop industry-based systems design cases, the author has used the following procedure. Completed and operational computer-based information systems for different industries were obtained and analyzed. The most important reporting from the systems, the necessary inputs, and the required processing were identified. The systems were reduced to the equivalent of "dots" by isolating the minimal information needed to understand the requirements of the system.

Given this definition ("dot equivalent") of a system, the student is asked to fill in the "lines" necessary to precisely specify the system. The systems are of necessity abbreviated, as are accounting "work sets," but the essence of the systems is captured in the case projects.

Contents of Cases

Each case is presented as a scenario in which the student is led through a series of events leading up to the specifications required to complete the case. The student is referred to in second person in the scenario. The standard contents of each case are outlined below.

1. *Industry Discussion*—At the first of each case, the student is given a brief explanation of the industry from which the case is derived.
2. *Introduction*—In the introduction, the organizational context and the student's role as a systems analyst are defined.
3. *Executive Meeting*—The systems analyst attends an executive meeting in which the organizational departments and the various managers are introduced. A major systems development project to be addressed with the assistance of a project team is revealed to the systems analyst.
4. *Project Team Meetings*—A series of project team meetings are conducted during which various managers present their information requirements. Definitions of reports, transactions, and processing requirements are stated in narrative form.
5. *Executive Presentation*—The results of the project team meetings are presented to top management. Top management adds a few addi-

tional reporting requirements and directs the systems analyst to prepare detailed systems specifications.

6. *Requirements to Complete Case*—The systems analyst is given the final instructions necessary to complete the specifications for the case.

The experience with this approach to systems analysis instruction has been favorable. The author and several colleagues have experimented extensively with a payroll system, the first case developed. This payroll case takes students an average of 10 to 12 hours to complete. It is included in this book (end of Chapter 1) as an optional training case that can be used before the major cases. The payroll case is not recommended for advanced or graduate students.

The book includes major cases in wholesale distribution, manufacturing, banking, and hospitals. Each case is designed as a major semester project requiring twenty-five to forty hours to complete. A case can be worked on by more than one student, but two students is the maximum recommended. NOTE: If time is a constraint, cases can be abbreviated by eliminating the reports requested during the executive presentation.

The availability of a repertoire of industries from which students can select cases is extremely valuable. It allows students not only to apply their technical skills, but also to become familiar with different industries. Accordingly, students can begin to determine areas of career interest and, due to their increased knowledge of an industry or industries, interview for jobs more effectively. That is, students learn the major information processing in an industry and the vocabulary associated with that information processing. They are, therefore, more comfortable and knowledgable during job interviews in that industry.

Student responses to the cases have been excellent. The students have been able to complete the cases, have increased their confidence in their technical skills, and frequently comment on the insight they have gained into the rigors required to adequately document systems specifications.

Acknowledgements

In conclusion, I wish to acknowledge the contributions of several individuals. I am especially indebted to V. Thomas Dock, University of Southern California, who encouraged me to write the book and served as consulting editor. The NCR Company gave invaluable assistance by allowing me to improvise from their *Accurately Defined Systems (ADS)* and information systems literature.

My colleagues at the University of Houston—Richard Scamell and Al Napier—graciously experimented with my cases as they were being developed. The students at the University of Houston patiently cooperated with rough working copies of the cases. The reviewers whose helpful suggestions are reflected in this book and in SYSTEMS ANALYSIS FOR COMPUTER-BASED INFORMATION SYSTEMS are: Marilyn Bohl, IBM;

Thomas I. M. Ho, Purdue University; Steve Alter, University of Southern California; Roger Hayen, University of Nebraska; M. Bond Wetherbe, Office of Information Systems, U.S. Naval Oceanographic Office; Charles Davis, MIS, Occidental Petroleum Corp.; Hank Lautenback, Idaho State University; and Charles Paddock and Phillip Judd, University of Houston.

I acknowledge A. Benton Cocanougher, Dean of the College of Business Administration, University of Houston, for his support of my efforts. Sincere thanks to three patient, tolerant, and forgiving secretaries: Cheryl Patrick, Dodie Taylor, and Susan Childs.

Finally, I acknowledge the efforts and support of my wife, Smoky, who is a supportive friend and colleague in my professional efforts.

As I acknowledge the efforts of all who have contributed, I also assume full responsibility for any inadequacies or discrepancies in the book.

Houston, Texas   **James C. Wetherbe**
1979

*Cases
in
Systems Design*

Chapter 1

Systems Specifications
Model for Systems Design

Systems Design and Specification
    Encoding and Compression
    Output Definitions
    Input Definitions
    Data Element Dictionary
    Decision Tables
    Systems Flowchart

Design Review

Summary

Exercises

Selected References

Payroll Case

This chapter presents the concepts and techniques that are the core (and perhaps the glamorous aspect) of systems analysis—the design of information systems. During the design stage, the systems analyst begins designing the means of solving the problems or capitalizing on the opportunities discovered during systems analysis. The systems analysis stage defines the way things are; the systems design stage defines the way things should be. The material covered in this chapter is preparation for selecting and completing one or more of the industry cases contained in Chapters 2 through 5.

SYSTEMS DESIGN AND SPECIFICATION

When structuring information for an information system, a systems analyst may not know in advance the specific hardware and software that will be best suited to support the system. Therefore, systems design techniques and procedures should be as hardware and software independent as possible. This allows the systems specifications to be implemented using whatever hardware and software turns out to be most cost effective in supporting the information structure. The relationship between information structure design and hardware and software is similar to the relationship between architectural design and building materials. In both instances, the specifications should articulate with sufficient precision what the final product should be without unduly restricting the selection of the best available means for constructing the final product.

Achieving a balance between design techniques that are both hardware/software independent and yet specific enough to be practical presents somewhat of a dilemma for systems analysts. In this book, an improvised and updated version of a technique entitled Accurately Defined Systems (ADS)[1] is used as a tool for designing specifications. This technique is a means of developing precise systems specifications without unduly imposing restrictions concerning particular hardware and software.

The process of structuring information involves the following specifications:

1. *Output Definitions*—Describe the printouts or terminal displays to be provided by the system.
2. *Input Definitions*—Display the format of each data field or element coming into the system.
3. *Data Element Dictionary*—Defines all data fields that are to be inputted, computed, stored, and reported.

---

1. Special appreciation is extended to the National Cash Register Company for granting permission for the use of material taken from *Accurately Defined Systems,* © copyright 1968, by the National Cash Register Company, Dayton, Ohio, U.S.A.

4. *Decision Table*—Illustrates the complex logical relationships necessary for input, computation, and reporting of information.

5. *Systems Flowchart*—Provides a graphical representation of the system, in which symbols represent operations, data flow, files, and equipment used.

The various design specifications are linked together such that the input, computation, logic, storage, and reporting of each item of information can be traced forward or backward from its point of origin to its final use(s). Upon completion, the specifications should provide sufficient documentation to guide the computer programming required.

Though some organizations have systems design procedures that differ from the procedures presented in this chapter, these differences are generally cosmetic rather than conceptual in nature. Readers who master the techniques presented in the remainder of this chapter should experience a comfortable transition to other systems design procedures.

Encoding and Compression

Before discussing the five types of systems design specifications, we need to understand how techniques of data encoding and data compression are used to save time and space in defining data elements, report headings, and the like.

The form in which data are used external to computers is generally more lengthy than necessary or desirable for machine processing. Reducing the number of characters or digits required to represent information reduces storage and processing costs. It also saves space on transaction documents and reports.

Encoding and compression are two basic approaches to represent information efficiently. *Encoding* consists of translating data from one form to another. For example, job titles can be coded as numbers. *Compression* consists of eliminating characters that are not essential to the information meaning. Abbreviation is the most common compression technique; it is effectively achieved by the removal of vowels from information or the truncation of trailing characters. Examples of encoding and compression are given in Tables 1-1 and 1-2, respectively.

Frequently, information must be added to a table of coded information. Therefore, it often is desirable to leave gaps between codes to facilitate the insertion of new information. For example, in Table 1-1, a new position of Senior Programmer can be coded as 105.

Table 1-1. *Encoding of Job Titles*

Job Title	Code
Programmer	100
Computer Operator	110
Systems Analyst	120
.	.
.	.
.	.
Accountant	250
Auditor	260
Controller	265

Encoding generally requires less space than compression. However, codes often have to be translated back to their original form to be useful external to the computer. Compressed data can generally be understood without translation.

Output Definitions

The output from an information system consists primarily of printed reports and/or displays on a CRT (cathode-ray tube) terminal. The types of information on the various reports and displays are determined by the analysis of information requirements and/or requests from information users.

The definition of output involves laying out the formats for the various reports and/or displays. A report layout form is used to define the specific location of each character to be printed (see Figure 1-1). Layout forms generally contain 132–160 horizontal print positions and 30–60 vertical print positions. The target hardware should be considered when designing each report or display. For example, if a report is to be printed on a 132-character printer, then not more than 132 horizontal print positions should

Table 1-2. *Compression of Merchandise Description*

Merchandise	Compressed
Flashlight	Flshlght
Battery	Bttry
Hammer	Hmmr
Charcoal Grill	Chrcl Grll

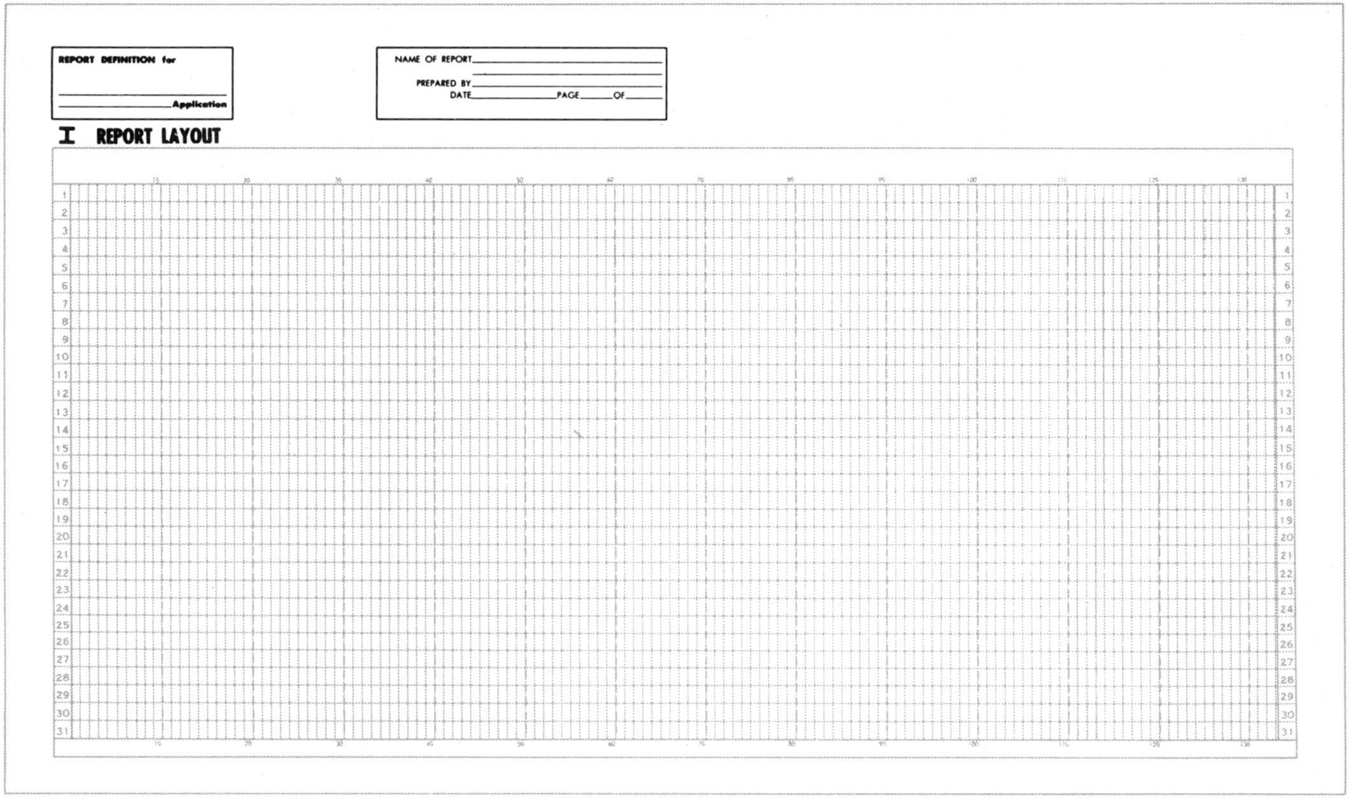

Figure 1-1. *Layout Form for Output Definition*

be used for one line. If output is to be displayed on a CRT, only the number of character positions available on that device should be used.

The actual format that should be used for a report or display is to a great extent a matter of taste. As a frame of reference, the systems analyst can review any existing reports or displays currently produced and/or ask persons who will use the reports or displays for their preferences as to output formats.

Headers and Information

Reports or displays contain both headers and information. *Headers* are the report titles, column headings, page numbers, and dates associated with a report or display. *Information* is the actual content of the output.

Headers may be printed by the computer or preprinted on custom-made forms. In either case, the headers are formatted on the layout form by positioning the letters of the headers using column and line numbers.

After the headers are formatted, the format of the information to be printed under the headers can be laid out. The symbols in Table 1-3 are used to convey, in a standard fashion, the formats of fields on the report. Information fields that are printed on successive lines need only have print symbols shown for the first line. A wavy vertical line drawn under a field is sufficient to convey that the field is to be repeated on successive lines, as illustrated in Figure 1-2.

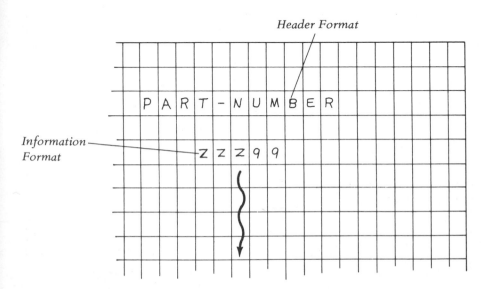

Figure 1-2. *Formatting Headers and Information*

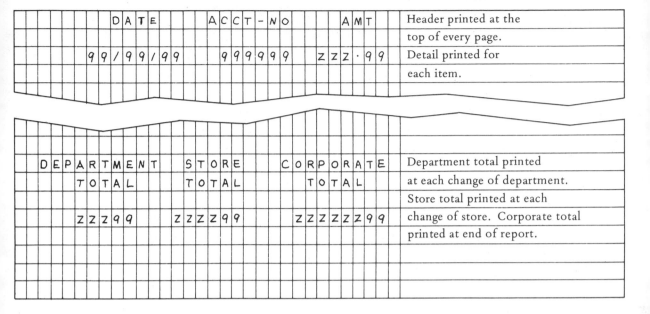

				D	A	T	E				A	C	C	T	-	N	O				A	M	T		Header printed at the
																									top of every page.
		9	9	/	9	9	/	9	9		9	9	9	9	9	9			Z	Z	Z	·	9	9	Detail printed for
																									each item.

	D	E	P	A	R	T	M	E	N	T		S	T	O	R	E		C	O	R	P	O	R	A	T	E	Department total printed
		T	O	T	A	L						T	O	T	A	L			T	O	T	A	L				at each change of department.
																										Store total printed at each
		Z	Z	Z	9	9		Z	Z	Z	Z	9	9				Z	Z	Z	Z	Z	Z	9	9		change of store. Corporate total
																										printed at end of report.

Figure 1-3. *Documenting Conditions under Which Report Lines Are to Be Printed*

The conditions causing each line of a report or a display to be printed should be conveyed on the report layout. The condition causing a line to be printed is generally a change in the values of certain information fields, or the start of a new page or display. For example, Figure 1-3 illustrates a report where totals are involved. The report is to have a detail header at the top of every page with successive lines of detail information below it. When the content of the department field changes, a department total is printed. It is reset to zero and then incremented again during the totaling of the detail for the next department. A store total is printed when there is a value change in the store field. Since the department field changes whenever the store field changes, it is also printed. Both department and store are then reset to zero and processing resumes. A corporate total is printed at the end of the report. It represents the grand total of all departments and stores.

Linking Output Definitions

To interface output definitions to the other design specifications requires some type of linking. The systems analyst must indicate the source of all information that is to appear on a report or display. To accomplish this, each field of information must be assigned a unique (and preferably meaningful) name. That unique name is then used to refer to that field of information or data throughout all design specifications. It is imperative that the systems analyst be consistent in both the unique identification and the spelling of each name. Otherwise, confusion and erroneous processing are likely.

Table 1-3. *Symbols Used to Format Information Fields*

9 = Numeric
X = Alphanumeric (characters or numbers)
Z = Zero suppress a numeric field
* = Lead numeric field for dollar protection
E = Embedded blank in an alphanumeric field
$ = Floating dollar sign if substituted for Z
· = Decimal point

Examples

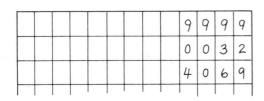

Four-digit numeric field with zero suppression. Any digit (0–9) may appear in any of the four positions.

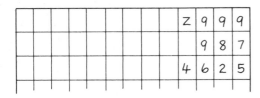

Four-digit numeric field with zero suppression for the first chatracter only. If the high-order position is zero, it appears as blank. All other digits are printed, regardless of value.

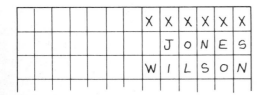

Six-character alphanumeric field. The digits 0–9, letters, and special symbols (%, *, ?, etc.) can appear in this field.

The names of the data fields shown on an output definition should be numbered and listed on a separate sheet of paper attached to the definition. The numbers of the names can then be placed within parentheses, directly under the corresponding data fields on the report or display layout. This provides a cross-reference between the fields printed on the report and the names of the fields (See Figure 1.4).

Though names act as a link to data fields on other design specifications, at this point the other specifications have not been prepared. Therefore, at the time that the report and display definitions are prepared, the source of a data field may not be known. This is appropriate; the output definitions are set up to determine what

Table 1-3. *(Cont'd)*

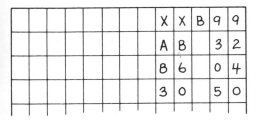

Five-position field containing two alphanumeric characters and two numeric characters, separated by a blank. The numeric characters are not zero suppressed.

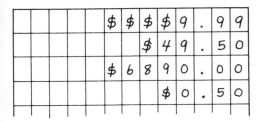

Eight-position numeric field with zero suppression on the first four positions. Multiple dollar signs indicate a floating dollar sign is to replace the right-most insignificant zero. All other insignificant zeroes are suppressed. The high-order digit of the output value should always be zero to insure the printing of a dollar sign.

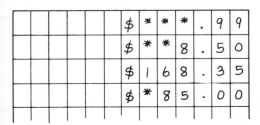

Seven-position field with a fixed dollar sign. The asterisk (*) indicates the use of dollar protection: leading asterisks are placed in unused dollar positions.

information is needed. Once the information needed is determined, the other specifications are used to determine what data and what processing are required to generate that information.

Sequencing and Selection

An output definition must also convey any sequencing or selection considerations of the output. *Sequencing* defines what, if any, sorting is needed prior to reporting. *Selection* defines what, if any, items are to be excluded from reporting. Both the sequencing and selection logic of a report can be documented at the bottom of the report layout or on a separate sheet attached to the report.

The sequencing of a report is expressed in terms of sequence

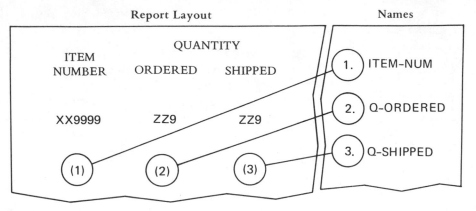

Figure 1-4. *Cross-Referencing Fields and Names*

levels. For example, if a report is to be sequenced by department number, it has one level of sequencing. If a report is to be sequenced by department number within store number, it has a two-level sequence. A sort on department number is required first; and it must be followed by a sort on store number. Generally, the highest, or last, level of the sort is called the *major sort*. Any additional lower levels are called *minor sorts*. A sorting sequence can then be expressed as follows:

Output Sequencing:

       Major Level 1. *Region Number*
            Level 2. *Store Number*
            Level 3. *Department Number*

In the preceding example, the output will be sorted on the last specified sort level (department number) first, then on store number, and finally on region number.

Selection logic is required when not all items of a given set of information are to be reported. In simple cases, the selection can be described narratively as follows:

Output Selection Logic:

    Exclude all items with amounts less than $1.00

For more complex selection logic, a decision table (discussed later in this chapter) should be used. When a decision table is used to specify the output selection logic, the output definition need only provide the number and/or the name of the decision table containing that logic.

Form Identification

One additional item of information necessary to complete an output definition is some type of form identification for documentation and linking purposes. Most layout forms provide some type of form identification (see Figure 1-1). The minimum points to be documented on each output definition are illustrated by the following example:

Name of Application or System:	Payroll System
Name of Report:	Payroll Checks
Prepared by:	John Smith
Date:	5/6/80

Input Definition

The input definition is used to lay out the format of all data coming into the system. Standard layout forms can be purchased or designed for this function, as illustrated in Figure 1-5. The form is not hardware-oriented. It can be used for any type of input medium.

Each input data field is assigned a unique name. This name is then used to refer to the data field throughout all design documentation.

The layout form defines the number of character positions in each field. The numbers at the bottom of each line show field positioning. Up to 80 character positions are available on one line of the form in Figure 1-5. Additional positions can be specified by continuing to the next line.

Record Length

Input transactions may be either fixed length or variable length. Under fixed length, each transaction has the same format and length as every other transaction. The weekly pay record in Figure 1-6 is a fixed-length transaction.

Variable-length transactions are most often used when a

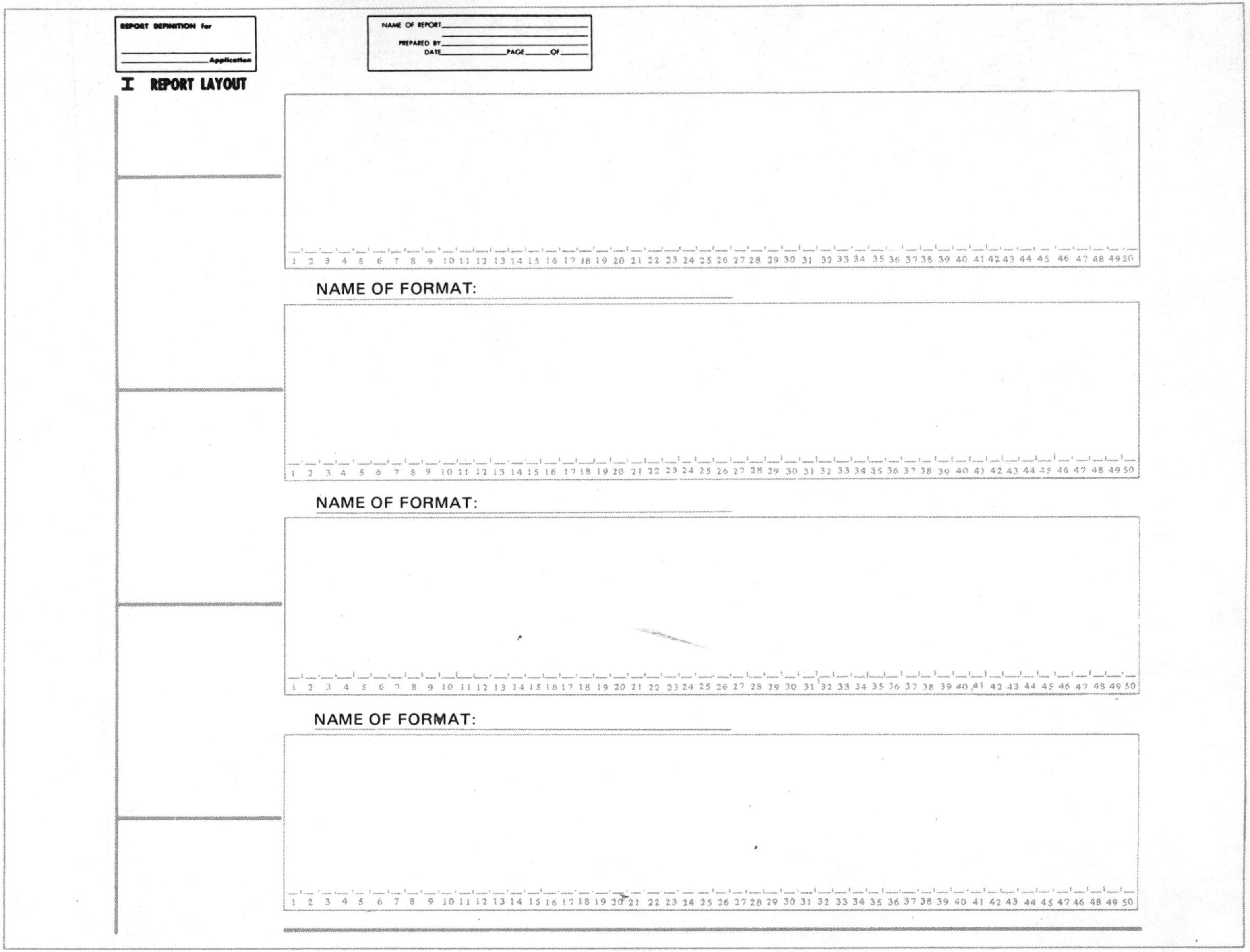

Figure 1-5. Layout Form for Input Definition

NAME OF FORMAT: Weekly-Pay

Figure 1-6. *Fixed-Length Transaction*

The fixed-length transaction contains the following fields across columns 1-50:

Columns	Field
1-6	DATE
7-16	SOCIAL SECURITY NUMBER
17-35	EMPLOYEE-NAME
36-37	HOURS-WORKED

NAME OF FORMAT: Registration

Figure 1-7. *Variable-Length Transaction*

The variable-length transaction contains the following fields:

Columns	Field
1-9	STUDENT IDENTIFICATION
10-30	STUDENT-NAME
31	CLASSIFICATION
32-33	MAJOR
34-35	RESIDENCY-CODE
36-37	NUMBER-COURSES
38-	COURSE-NUMBER (Repeats up to ten times, depending on number-courses)

field or group of fields may be repeated within a record. For example, when students enroll for classes, the number of courses they register for varies. Accordingly, for each transaction, there are fixed fields identifying a student in a standard format, followed by variable fields that are repeated for each course for which the student is registered (see Figure 1-7). The number of courses a student is enrolled for is indicated by the **NUMBER-COURSES** field that precedes the repeatable **COURSE-NUMBER** field. Notice that the layout of a repetitive field is shown one time. The repetition is conveyed by inserting the word *repeats* or a similar comment next to the field.

Multiple Inputs

Many applications accept more than one type of input transaction. For example, a payroll system might accept the following transactions:

1. New employee transaction—Indicates employee social security number, name, department, pay rate, date-of-hire, etc.
2. Weekly pay transaction—Indicates employee social security number, hours worked, date, etc.

The social security number is used in both transactions as the *key,* or record identification, to access the corresponding employee record in the payroll master file. However, whether a transaction is a "new employee transaction" or a "weekly pay transaction" needs to be identifiable to the system so that the transaction can be processed correctly. This is accomplished by the use of a *transaction code* that uniquely defines each type of input transaction. The transaction code can be located in any position of the transaction, but, once a position has been determined, it must be used consistently for all transactions inputted to a system. This is necessary since the computer program will first check this code to determine what processing is required for a particular transaction. In the case of the two types of payroll transactions, the transaction code can be located in the last two columns of the transaction. The code values 01 and 02 can stand for new employee and weekly pay transactions, respectively.

Linking Input Definition

An input definition needs to be linked to both the source document from which input data are collected and the remainder of the systems design specifications.

The input definition defines the format in which data are to come into the computer system. However, it is highly unlikely that this format will be used for original data capture. Rather, a more familiar and traditional transaction document format such as a customer order form or bank deposit slip is generally used. The contents of these documents are then converted into machine-readable form in the format of the input definition. Accurate cross-referencing of the data fields from the original (source) document to the input definition is therefore required. This can be accomplished using a coding scheme similar to the one described for an output definition in which corresponding data fields are numbered identically.

The input definition is linked to the remainder of the systems specifications by the use of a unique name for each data field. As noted earlier, each data field should be referred to in exactly the same way throughout all design documentation.

Form Identification

Each input definition must be identified for documentation purposes. An area for this identification is usually provided on the input definition form (see Figure 1-5). The minimum points to be documented on each input definition are illustrated by the following example:

Name of Application or System:	Payroll System
Name of Transaction:	Weekly Pay Transaction
Sequence of Transactions:	Social Security Number
Media Type:	Punched Card
Prepared by:	John Smith
Date:	5/6/80

Data Element Dictionary

The data element dictionary defines every data field in the system. The data element dictionary provides the following information for each data element:

> *Name*
> *Definition*
> *Source*
> *Where Used*
> *Maintenance*
> *Storage*

Name

The data element name is the unique name assigned to the data element. This name can be used to track an element referred to in an output, input, or decision table back to the data element dictionary for a complete description.

Because the data elements will be referred to in computer programs, it is helpful to choose data names that can be used as names in computer programs. This allows data names in computer programs to be directly cross-referenced to the data element dictionary. Programming languages such as COBOL and PL/1 allow the use of lengthy (generally a minimum of 25 characters) and meaningful data names. When appropriate, characters within names can be separated by hyphens or underscores for clarity. RPG, another commonly used programming language, generally allows only six characters per name. By using the compression techniques discussed earlier in this chapter, the systems analyst can create data names that can be used in computer programs. For example, a social security number field can have one of the following data names:

SOC-SEC-NUM

SS-NUM

SSN

Such abbreviations save time and space in coding specifications and programs. However, an abbreviation should not be carried to the point that it becomes difficult to tell the original name of the data element. If the programming language to be used is not known when the data names are created, the names can be modified if necessary after the language has been determined.

Definition

The definition provides both a field format and a narrative description of the meaning of the variable. For example, QUANTITY-ON-HAND can be defined as the number of items in existing inventory excluding any back orders and any items committed to customers but not yet shipped.

If a data element is to be expressed as a code, the code values and their meanings should be included in the definition. For example, the possible code values for sex may be stated as follows:

Code	Meaning
1	Male
2	Female

The format of a field is defined by indicating both the length of the field and whether the data to be stored in the field are numeric or alphanumeric. "9's" and "X's" are used to represent numeric and alphanumeric characters, respectively. For numeric fields, the allowable sign values (i.e., + and/or −) should be indicated. If a decimal point is involved, its position should be defined.

Source

Data elements can come from one of three places: inputs, tables, or computations. Inputted data elements come directly into the system from an input data stream (e.g., a transaction file). Table data elements are looked up in predefined tables (e.g., tax rate tables). Computed data elements are computed from other data elements in the system.

For a data element that is inputted, the input source is identified and any input validation rules are defined. For example, validation rules for the date field of a transaction follow:

Validation Rules:
    Cannot be later than today's date or 30 days earlier than today's date.

For a data element to be determined by a table lookup, the table values and the indexes used for table lookups are defined. This can be accomplished by simply displaying the table. For example, a **SALARY** table with indexes **PAY-GRADE** and **PAY-STEP** is shown below:

SALARY Table:

			PAY-STEP		
PAY-GRADE	1	2	3	4	5
1	500	532	568	612	688
2	645	720	785	822	876
3	832	912	986	1666	1188
4	1112	1186	1290	1380	1440
5	1360	1502	1612	1720	1816
6	1706	1820	1920	2012	2118

PAY-GRADE and **PAY-STEP** are data elements that must be defined elsewhere in the data element dictionary. The **SALARY**

table is used by locating the point of intersection for a given set of values for PAY-GRADE and PAY-STEP. For example, a PAY-GRADE of 4 and PAY-STEP of 3 point to a SALARY table value of 1290.

Table values may be maintained external to the program (e.g., on cards or disk) and inputted during program execution. Alternatively, they may be coded as part of the program. If table values change frequently, it is preferable to maintain them external to the program so that they can be changed without modifying program code.

For a data element that is computed, the necessary computation is defined. This can be done in simple algebraic form with one minor variation. The equal sign (=) in computer processing means that the variable on the left side of the equation is "set equal to" the computation or variable on the right side of the equation. For example, if A = 6 and B = 3, the equation C = A + B + 2 results in C being set equal to 11, irrespective of C's value prior to the computation.

Each of the data elements used to compute the value for a computed data element must be defined elsewhere in the data element dictionary. For cross-reference purposes, care must be taken to insure consistent spelling of the name of each data element.

If a sequence of computations is necessary to compute a data element, the complete sequence must be defined. For example, the following sequence of computations may be used to compute BALANCE-DUE.

Computation:

1. PAYMENT = CASH-DEPOSIT + CHECK-DEPOSIT
2. NEW-BALANCE = OLD-BALANCE − PAYMENT
3. INTEREST = NEW-BALANCE × INTEREST-RATE
4. BALANCE-DUE = NEW-BALANCE + INTEREST

Logical processes may be required when inputting or computing data elements. In a simple case, the logical process may be described in the data element dictionary. For example, consider the following logic associated with the computation of overtime:

IF HOURS WORKED IS GREATER THAN 40 DO COMPUTATION 2
OTHERWISE DO COMPUTATION 1

1. GROSS-PAY = HOURS-WORKED × PAY-RATE
2. OVERTIME = HOURS-WORKED − 40
3. GROSS-PAY = (40 × PAY-RATE) + (OVERTIME × (1.5 × PAY-RATE))

For complex logical relationships, decision tables (discussed in the following section) should be used. Decision tables are assigned numbers and/or names so they can be referenced from the data element dictionary as follows:

Computation: Refer to Decision Table 1, **OVERTIME-COMPUTATION**, to determine which of the following computations to use:

1. Computation 1
2. Computation 2
          :
          :

n. Computation n

Where Used

A where-used list is created to keep track of each report and computation in which a data element is used. Such a list is needed for control purposes.

Consider a situation where the data element **SALARY** is printed on several reports and displays available throughout the organization. If the organization (or the government) decides distribution of this salary information should be restricted, the where-used list pinpoints the reports and displays what should be reviewed.

In another situation, the definition of a data element may be changed. For example, assume the data element **QUANTITY-ON-HAND** is redefined to include back orders. In such a case, it is judicious to review all computations in which **QUANTITY-ON-HAND** is used to check for unanticipated effects on computations. Also, the individuals who receive reports affected by this change should be identified and alerted to the new procedure.

Maintenance

Maintenance information about a data element describes the procedures for maintaining or updating a data element. The concept of documenting maintenance is best illustrated by examples:

- **SEX-CODE:** Originated when employee is first put on the payroll. No maintenance is required unless an error is detected.
- **BALANCE-DUE:** Computed during the billing cycle for each customer in the accounts receivable file.
- **SALARY:** Updated after each annual review or when a merit increase is granted.

Storage

Storage defines if and where (i.e., in what file) a data element is to be stored. Not all data elements are stored in a file. For example, in a weekly payroll for hourly employees, GROSS-PAY is computed by multiplying PAY-RATE by HOURS-WORKED. Since HOURS-WORKED is inputted each week and GROSS-PAY is computed each week, neither data element has to be stored in the payroll file. In contrast, PAY-RATE changes only when a raise is given. It is relatively stable from pay period to pay period. Consequently, PAY-RATE should be stored so that it is conveniently available for payroll processing.

Storage for GROSS-PAY and HOURS-WORKED is defined as follows:

> *Storage:*   Not Applicable (or N/A)

Storage for PAY-RATE is defined as follows:

> *Storage:*   Payroll Master File

At the time the data element dictionary is defined, whether or not certain data elements are to be stored may not be determined. In such cases, the storage definitions may remain blank until a decision has been made.

Organization

There is no standard form for a data element dictionary. Since the amount of documentation required for a data element varies, defining a data element is best accomplished on blank paper or index cards using the definition categories in an outline format. Figure 1-8 shows how a completed entry in the data element dictionary might appear.

For reference purposes, it is convenient to maintain the data element dictionary in alphabetical order. This may be difficult if unanticipated data elements are added frequently during design. However, if each data element is documented on a separate sheet of paper or index card, insertion (and deletion) of elements is accomplished easily. It may be possible to use an online system when creating and maintaining the data element dictionary. This is a desirable enhancement, but not a necessary one.

After all data elements have been defined, they can be organized into one or more files. This is accomplished by clustering related data elements into groupings. For example, in an order processing system, the data element dictionary contains data elements relative to both customers and inventory. This leads

Name	DEPARTMENT
Definition	The code of the department an employee works in. Format is 999
	Code values are as follows: 101–Administration 201–Production 301–Sales 401–Engineering
Source	Input: New employee transaction Validation rule: Code value must be equal to a valid department code.
Where Used	Payroll Check Validation Error Report Department Payroll Report
Maintenance	Originated from new employee transaction when employee is hired. Updated whenever an employee changes departments.
Storage	Payroll Master File

Figure 1-8. *Documentation for a Data Element*

to the creation of a customer master file and an inventory master file.

Decision Tables

Decision tables are powerful, efficient tools for expressing complex logical relationships in an understandable manner. Decision tables may be used in conjunction with the data element dictionary and output definitions as follows:

1. They may be used in conjunction with the "source" section of the data element dictionary to define logic used in input validation or computations.
2. They may be used in conjunction with the "output selection logic" section of the output definition to define logic used for selecting data to be included in a report or terminal display.

Terminology and Structure

A decision table is a matrix containing columns and rows that are used to define relationships. Figure 1-9 illustrates a decision-table format.

There are three major components in a decision table: con-

DECISION TABLE FOR _____ APPLICATION

DECISION TABLE NAME _____ REFERENCE NO. _____

PREPARED BY _____ DATE _____

CONDITIONS/
COURSES OF
ACTION

DECISION RULES

01	02	03	04	05	06	07	08	09	10	11	12	13	14	15	16	17	18	19	20	21	22	23	24	25	26	27	28	29	30	31	32	33	34	35	36	37	38	39	40	41	42	43	44	45	46	47	48	49	50

Figure 1-9. *Layout Form for a Decision Table*

ditions, courses of action, and decision rules. *Conditions* are events or facts that determine the courses of action to be taken. *Courses of action* are processes or operations to be performed under certain conditions. *Decision rules* express the relationships between combinations of conditions and courses of action.

Decision rules are expressed by making condition entries and course-of-action entries in the matrix provided to the right of condition statements and course-of-action statements (see Figure 1-9). Possible decision-rule entries and their definitions follow:

1. Condition Entries
   Y = Yes; condition statement must apply.
   N = No; condition statement must not apply.
   – = Indifferent; condition statement is irrelevant to the decision and does not have to be considered.
2. Course-of-Action Entries
   X = Activate course of action.
   · = Do not activate course of action.

Illustration of a Decision Table

Figure 1-10 illustrates a decision table that defines customer billing for the output selection logic of an output definition. Note that a double horizontal line is used to separate conditions from courses of action.

If a customer's balance is less than or equal to zero (decision rule 1, read vertically), no additional conditions have to be checked. A statement is not to be printed for the customer. If a customer's balance is greater than zero, but less than his or her credit limit (decision rule 2), no additional conditions have to be

Figure 1-10. *Completed Decision Table*

CONDITIONS/ COURSES OF ACTION	DECISION RULES											
	1	2	3	4	5	6	7	8	9	10	11	12
BALANCE–DUE ≤ 0?	Y	N	N	N								
BALANCE–DUE > 0 AND < CREDIT–LIMIT?	–	Y	N	N								
BALANCE–DUE > CREDIT–LIMIT?	–	–	Y	Y								
CREDIT RATING EXCELLENT?	–	–	Y	N								
DO NOT PRINT STATEMENT	X	·	·	·								
PRINT STATEMENT	·	X	X	X								
ADD CREDIT WARNING TO STATEMENT	·	·	·	X								

checked. A statement is to be printed for the customer. If the customer's balance exceeds his or her credit limit, then the credit rating is checked. If the credit rating is excellent (decision rule 3), a statement is to be printed for the customer. If the credit rating is not excellent (decision rule 4), a statement is to be printed, and a credit warning message is to be added to it.

Constructing Decision Tables

The approach used to construct decision tables varies among systems analysts. The technique discussed below is based on the *progressive rule development* approach advocated by Keith London.[2] This technique is easy to understand and keeps the logic of the problem in sight at all times.

The formal rules for constructing a decision table using progressive rule development follow:

1. Conditions and courses of action should be listed in the sequence in which they are to be considered.
2. Consider a condition (relevant to the decision) in the positive case (Yes).
3. Consider, in the positive case, all other conditions (relevant to the decision) that must be considered before a course of action can be taken.
4. Use the indifference code ("−") for any conditions that are irrelevant to the decision rule under consideration.
5. Enter the conditions, courses of actions, and their respective decision-rule entries in the decision table.
6. Start a new decision rule by negating the last positive condition in the current decision rule (leave all entries the same).
7. Repeat Step 3 onward until the table is complete.

The decision rules in Figure 1-10 were determined using progressive rule development. Review the rules in Figure 1-10 (sequentially, 1 through 4) in the context of progressive rule development.

Two additional examples of decision tables are provided in Figures 1-11 and 1-12. Note the use of progressive rule development in both examples.

Figure 1-12, a decision table for airline reservations, introduces a new concept. The condition "First-class available" is stated as the third and sixth conditions. This repetition is required

2. London, Keith R. *Decision Tables*, Auerbach, 1972.

CONDITIONS/ COURSES OF ACTION	DECISION RULES									
	1	2	3	4	5	6	7	8	9	10
Salaried?	Y	Y	N	N	N					
Hourly?	—	—	Y	Y	Y					
Hours worked < 40?	Y	N	Y	N	N					
Hours worked = 40?	—	—	—	Y	N					
Hours worked > 40?	—	—	—	—	Y					
Pay base salary	X	X	•	•	•					
Calculate hourly wage	•	•	X	X	X					
Calculate overtime	•	•	•	•	X					
Produce Absence Report	X	•	X	•	•					

Figure 1-11. Decision Table for Payroll Calculations

to adhere to the rule of listing entries in the sequence in which they are considered. In decision rules 5 through 8, the passenger is requesting economy-class. Therefore, there is indifference about first-class availability unless it is determined that no economy seats are available and an alternative is acceptable. At that point, the next condition to be considered is the availability of a first-class seat.

CONDITIONS/ COURSES OF ACTION	DECISION RULES							
	1	2	3	4	5	6	7	8
First-class request?	Y	Y	Y	Y	N	N	N	N
Economy-class request?	—	—	—	—	Y	Y	Y	Y
First-class available?	Y	N	N	N	—	—	—	—
Economy available?	—	Y	Y	N	Y	N	N	N
Alternative acceptable?	—	Y	N	—	—	Y	Y	N
First-class available?	—	—	—	—	—	Y	N	—
Reduce first-class available	X	•	•	•	•	X	•	•
Issue first-class ticket	X	•	•	•	•	X	•	•
Reduce economy available	•	X	•	•	X	•	•	•
Issue economy ticket	•	X	•	•	X	•	•	•
Refer to alternate flight	•	•	X	X	•	•	X	X

Figure 1-12. Decision Table for Airline Reservations

The repetition of condition statements or course-of-action statements is occasionally necessary to insure clarity when using progressive rule development. This type of presentation is helpful to computer programmers because the logic can be conveniently translated into program code in the same sequence as it is presented in the decision table.

Form Identification

Each decision table should be identified for documentation purposes, preferably in an area provided on the decision table form. The minimum points to be documented are illustrated in the following example:

Name of Application or System:	Payroll System
Decision Table Name:	Pay Calculations
Reference Number:	112
Prepared by:	Jill Smith
Date:	5/30/80

Systems Flowchart

Systems specifications are completed by preparing a *systems flowchart*. The flowchart is a graphical representation of the system. Symbols are used to represent operations, data flow, files, and equipment used. Figure 1-13 shows the symbols that can be used in constructing a systems flowchart. *Flowchart templates*[3] can be purchased and used to accurately form the various flowchart symbols.

Figure 1-14 shows an order processing system. Note that the order processing system uses the inventory master file in the processing of orders. However, the order processing system does not include all the necessary inventory processing. Another subsystem is required for maintaining (adding, deleting, and modifying) inventory data. This subsystem is shown in Figure 1-15. The inventory master file used in the inventory subsystem is the inventory master file used in the order processing subsystem; that is, the identical symbols on the flowcharts represent the same file.

---

3. Flowchart templates are thin plastic plates with cutout forms of flowchart symbols.

Figure 1-13. *Flowchart Symbols*

Processing A major processing function.	**Input/** **Output** Any type of medium or data.
Punched **Card** All varieties of punched cards, including stubs.	**Perforated** **Tape** Paper or plastic, chad or chadless.
Document Paper documents and reports of all varieties.	**Transmittal Tape** A proof or adding-machine tape or similar batch-control information.
Magnetic **Tape**	**Disk, Drum,** **Random Access**
Offline **Storage** Offline storage of either paper, cards, magnetic tape, or perforated tape.	**Display** Information displayed by plotters of visual–display devices.
Online **Keyboard** Information supplied to or by a computer utilizing an online device.	**Sorting,** **Collating** An operation on sorting or collating equipment.
Keying **Operation** An operation utilizing a key-driven device.	**Clerical** **Operation** A manual offline operation not requiring mechanical aid.
Auxiliary **Operation** A machine operation supplement- ing the main processing function.	**Communication** **Link** The automatic transmission of information from one location to another via communication lines.

Flow

The direction of processing or
data flow.

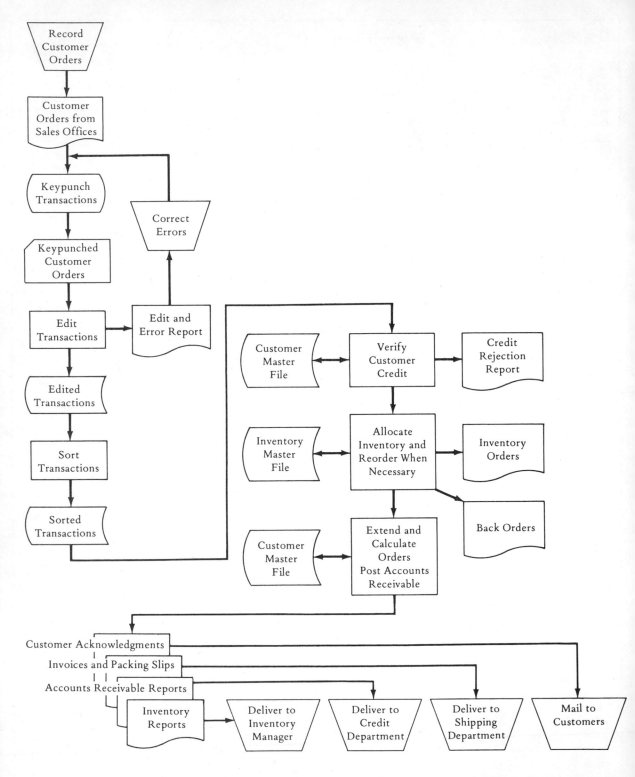

Figure 1-14. *Flowchart of an Order Processing Subsystem*

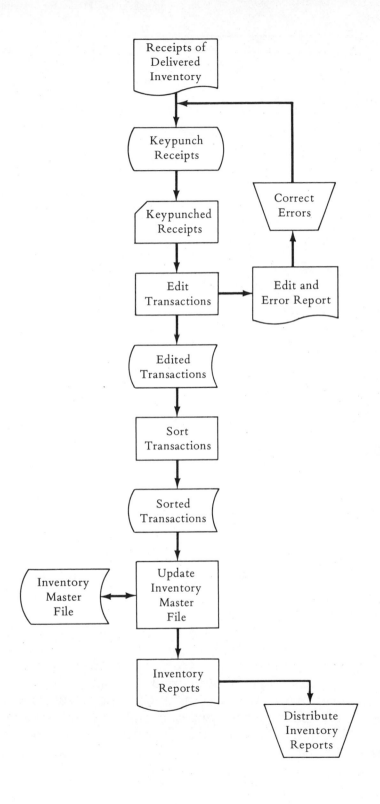

Figure 1-15. *Flowchart of an Inventory Subsystem*

INVENTORY MASTER FILE	CUSTOMER MASTER FILE
ITEM-NUMBER	CUSTOMER-NUMBER
DESCRIPTION	CUSTOMER-NAME
CLASS	CUSTOMER-ADDRESS
VENDOR-NUMBER	ZIP-CODE
ITEM-WEIGHT	PHONE-NUMBER
OUT-OF-STOCK-CODE	YEAR-ACCT-OPENED
TAX-CLASS	CREDIT-LIMIT
UNIT-OF-MEASURE-CODE	CREDIT-RATING
UNIT-OF-MEASURE	CREDIT-RATING-DATE
SALESPERSON-COMMISSION-RATE	SALESPERSON-NUMBER
SUGGESTED-RETAIL-PRICE	SERVICING-BRANCH-CODE
PRICE-CODE	DATE-OF-LAST-SALE
PRICE-A	CURRENT-BALANCE
QUANTITY-BREAK-A	
PRICE-B	
QUANTITY-BREAK-B	
PRICE-C	
QUANTITY-BREAK-C	

Figure 1-16. File Contents in an Order Processing System

Constructing a Flowchart

Since the other systems specifications describe the inputs, processes, files, and outputs of the system, the systems flowchart need only link these various systems components. Each input document or display and each output document or display on the flowchart should be labeled exactly as its respective input or output definition is labeled.

Data elements can be organized into one or more files and given file names. Figure 1-16 lists the contents of the inventory and customer master files used in an order processing system. The computer programs needed to perform the input, processing, and output functions can then be determined, given names, and charted into the system. Computer programs are required between the following components:

Components	*Program Function*
Inputs and outputs	List input transactions
Inputs and files	Update file(s) with transactions
Files and outputs	Create report(s) using file(s)
Files and files	Merge two files into one file

Documentation of any required data preparation and report distribution steps completes the flowchart.

DESIGN REVIEW

Usually, several persons are involved and several months elapse in preparing systems specifications. Due to the number of people involved, the amount of time required, and the complexity inherent in developing systems specifications, there are frequently omissions, inconsistencies, and other errors that are not recognized until the various specifications are brought together.

An effective technique to highlight any problems within the systems specifications is a design review. A *design review* is a formal presentation of the systems specifications. It is a "guided tour" of the system by those who developed the specifications.

The audience for the design review may consist of interested users and/or managers and, of course, members of the systems design team. The design review is a learning experience for all involved.

Using the system's flowchart as the nucleus, a design review draws from the input and output specifications, data element dictionary, and decision tables to provide a detailed review of systems functions and capabilities. The design review is an excellent tool for uncovering errors and eliminating them. It serves as a positive motivation for those involved in preparing systems specifications and provides a measure of completeness before proceeding with systems development.

SUMMARY

Systems design is involved with determining what information to include in an information system and the structure of it. The systems analyst plays the lead role in the structuring of information to effectively and efficiently support the organization's information requirements.

The systems analyst, in conjunction with management, develops the following systems specifications:

Output Definitions
Input Definitions
Data Element Dictionary
Decision Tables
Systems Flowcharts

The output definitions describe printouts and terminal displays. The input definitions define the format of each data element coming into the system. The data element dictionary defines all data fields that are inputted, computed, stored, and reported. Decision tables illustrate complex logical relationships. The systems flowchart provides a graphical representation of the system.

These specifications are linked together such that the input, computations, logic, storage, and reporting of each data element can be traced forward or backward from its point of origin to its final use(s). The completed systems specifications provide sufficient documentation for the computer programming of the system.

EXERCISES

1. Illustrate how to define a six-character numeric data field that is to have a decimal point between the hundreds and tens positions. How many positions on an output definition does this field require? How would the field appear with field values of 30? 488? 5069?

2. Construct an output definition field format for inventory part codes to appear as AE91609; CG41113; FR60021.

3. How is an item in an output definition linked to the data element dictionary?

4. Create an input definition format for the following input transaction:
    *   The first field is a nine-digit account number.
    *   The second field is a fifteen-character field for customer name.
    *   The third field is a six-digit date field in the format of month, day, year.
    *   The fourth field is a two-digit transaction code.
    *   The fifth field is a six-digit field called AMOUNT.

5. Where does a system analyst define the computations and/or the reports in which a data element is used?

6. Why is it essential to assign unique names to data elements and always use their correct spellings?

7. What are the three possible sources of data elements in an information system?

8. Why is it important to document where data elements are used?

9. What determines whether a data element should be stored in a file?

10. Use decision tables to illustrate the logic for the following situations:
   a. ABC Inc. must deduct city tax from employee wages depending upon where they live and work as follows:
      1. Employees living in the city must pay the tax.
      2. Employees living outside the city and working within the city must pay the tax.
      3. Employees living and working outside the city do not pay the tax.
   b. Salespersons for XYZ Distributing have the possibility of receiving a monthly bonus check. To receive a bonus, a salesperson must achieve the following goals:
      1. He or she must have sold in excess of ten units or sold orders to a minimum of two new customers.
      2. He or she must have installed three units or sold an additional five units or sold an additional unit to a new customer.
      3. He or she must have turned in all expenses for the past month and have submitted reports on contacts of at least five new prospects.

SELECTED REFERENCES

London, Keith R. *Decision Tables*. Pennsauken, N.J.: Auerbach, 1972.

Lucas, Henry C. *The Analysis, Design, and Implementation of Information Systems*. New York: McGraw-Hill, 1976.

Montalbano, Michael. *Decision Tables*. Palo Alto: SRA, 1974.

National Cash Register Company. *Accurately Defined Systems*. Dayton, Ohio, 1968.

Nolan, Richard L. "Systems Analysis for Computer Based Information Systems Design," *Data Base*, Winter 1971, pp. 1–10.

Semprivivo, Phillip C. *Systems Analysis: Definition, Process, and Design*. Palo Alto: SRA, 1976.

Schussel, George. "The Role of the Data Dictionary," *Datamation*, June 1977.

Wetherbe, James C. "A Systems Specification Model for Instruction in Systems Analysis and Design," *Journal of Data Education*, July 1978.

Wetherbe, James C. "Development and Application of Industry-Based Cases in Systems Analysis and Design," *Journal of Data Education*, October 1978.

Payroll Case

This case study applies the systems design and specifications concepts and techniques discussed in Chapter 1. The system to be designed is a payroll system. A payroll system is an information system common to all organizations.

The case is provided as an optional training exercise to precede the major cases contained in Chapters 2 through 5. The case is not recommended for advanced or graduate students.

The case is structured to be realistic, but it is abbreviated so that it can be completed in a short period of time. There are many ways the case can be expanded into a major time-consuming project. To avoid doing this, you should stay within the requirements defined in the case.

INFORMATION REQUIRED FROM THE SYSTEM

The following basic reports are necessary for decision making and/or for documenting transactions (e.g., for auditing purposes).

Pay Statement (See Figure 1.)

1. Employee Number (EMP-NO)
   A six-character, numeric field with no zero suppression.
2. Employee Name (NAME)
   A twenty-character, alphanumeric field.
3. Social Security Number (SOC-SEC-NO)
   A nine-character, numeric field with hyphens between the fourth and fifth positions from the right and the sixth and seventh positions from the right.
4. Hours Worked (HRS)
   A four-character, numeric field. It is to have a decimal point between the tens and units positions and zero suppression to the decimal point. This is the total hours worked for the week.
5. Department (DEPT)
   A five-character, numeric field with zero suppression in the

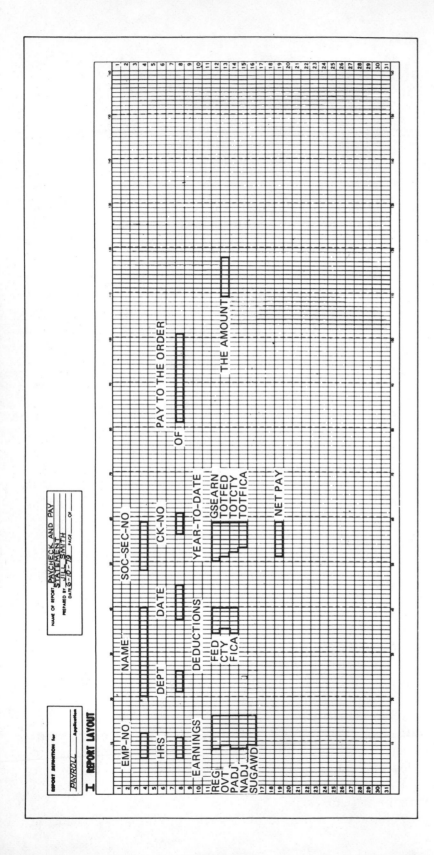

Figure 1. *Format of Paycheck and Pay Statement*

two high-order positions. The employee will not work in more than one department per week. If there is a departmental transfer within the system, it will be picked up the following week through a change record. The department numbers range from 100 to 99999.

6. **Date Check Is Written (DATE)**
A six-character, numeric field with month, day, year format. The month and day fields should have the possibility of zero suppression in their high-order positions. There should be slashes (/) between the month and day, and the day and year. This is today's date (actual date). It will be available within the computer.

7. **Check Number (CK-NO)**
A five-character, numeric field with zero suppression up to but not including the units position. A consecutive check number is to be printed on each check. The range is from 1 to 99999. When the upper limit is reached, this field should be reset to 1.

8. **Earnings**
The fields under this heading represent all possible ways to earn income. All fields are to have decimal points between the hundreds and the tens positions.

9. **Regular Earnings (REG)**
A seven-character, numeric field with zero suppression up to the decimal point. Regular earnings is basic pay less any overtime income, which is to be itemized separately on the next line of print. A salaried employee's wages are considered regular earnings. Regular earnings for an hourly worker are hours ($\leq$40) times the rate. The employee will have only one rate per week.

10. **Overtime Earnings (OVT)**
A six-character, numeric field with zero suppression up to the decimal point. Overtime earnings can be received only by an hourly employee. Overtime is based on hours over 40. All hours worked over 40 are considered overtime income hours. The overtime rate is 1.5 of the regular rate for each hour of overtime.

11. **Positive Pay Adjustment (PADJ)**
A seven-character, numeric field with zero suppression up to the decimal point. There can be only one positive pay adjustment per week per employee. Any employee may have a positive pay adjustment.

12. **Negative Pay Adjustment (NADJ)**
A seven-character, numeric field with zero suppression up to

the decimal point. There can be only one negative pay adjustment per week per employee. Any employee may have a negative pay adjustment. It will never exceed the net pay for the week. There will be approximately ten to twenty of these per pay period. (An employee can receive both a positive and a negative pay adjustment on the same check.)

13. Suggestion Award (SUGAWD)

A six-character, numeric field with zero suppression up to the hundreds position (dollar position). All employees are eligible for a suggestion award. The minimum award is $1.00, and the maximum award is $6000.00. An individual may have more than one award per week. If he or she has more than one, the field on the check is to be a summation (total) of the awards. NOTE: The identifying captions REG, OVT, PADJ, FED, GSEARN, and the like, are not preprinted. If no data entry is to be made in a field, do not print the caption. Do not leave a blank line of print; shift the next entry up in its place. If no entries remain for a column, proceed to next print step. Except for these captions under Earnings, Deductions, and Year-to-Date, the rest of the check is preprinted.

14. Deductions

The following three fields represent the possible deductions. Each field is to have a decimal point between the hundreds and the tens positions, and each is to have zero suppression to the decimal point.

15. Federal Withholding Tax (FED)

A five-character, numeric field. This tax is to be deducted for all employees. The tax is a percentage of gross income for the week. This percentage is given in Table 1. It is based on the projected yearly earnings and the number of dependents claimed by the employee. (Projected yearly earnings are the total weekly earnings times 52.)

16. City Tax (CTY)

A four-character, numeric field. This tax is 1 percent of the employee's total earnings for the week.

17. Federal Insurance Contribution Act (FICA)

A five-character, numeric field. FICA (social security tax) applies to all employees. It is 6.05 percent of the gross income up to $17,700 for the year. (This formula is subject to change.) When the individual's earnings reach the $17,700 limit, this deduction is to cease for the rest of the current calendar year. NOTE: Do not overdeduct.

18. Year-to-Date

The following fields are to be the current status of these

Table 1. *Federal Withholding Tax Table Deduction Percentage*

Projected Yearly Earnings	DEPENDENTS						
	0	1	2	3	4	5	6 & up
Less than $6,000	0	0	0	0	0	0	0
$6,001 — $8,000	16	18	16	14	13	11	9
$8,001 — $12,000	18	16	14	12	11	18	9
$12,001 — $15,000	22	20	18	16	15	14	13
$15,001 — $20,000	24	22	20	19	18	17	16
$20,001 — $35,000	28	26	24	23	22	21	20
$35,001 — $60,000	32	30	29	28	27	26	25
Greater than $60,000	36	34	32	30	29	28	27

categories as of this pay period. All fields should have decimal points between the hundreds and the tens positions with zero suppression to the decimal point.

19. Gross Earnings (**GSEARN**)
   An eight-character, numeric field that is a total of all gross earnings of each week (year-to-date), including the current week.

20. Total Federal Withholding Tax (**TOTFED**)
   A seven-character, numeric field that is the total of all withholdings of this week's tax, year-to-date.

21. Total City Tax (**TOTCTY**)
   A seven-character, numeric field that is a total of all deductions made each week (year-to-date) for city tax. The same rules apply here as when it is deducted.

22. Total FICA (**TOTFICA**)
   A five-character, numeric field that is an accumulation of this deduction year-to-date. It will become a fixed figure when the upper limit is reached.

23. Net Pay (**NET PAY**)
   A seven-character numeric field with a decimal point between the hundreds and the tens positions with zero suppression up to the decimal point. Net pay is the total earnings for the week minus the total deductions for the week. This amount will never be negative.

Paycheck (See Figure 1.)

1. Employee Name
   The name field (Pay to the Order of) is a twenty-character, alphabetic field. This name is to appear on the statement (STUB).
2. A seven-character, numeric field with a floating dollar sign to the decimal point. The dollar sign is to be displayed in the first character position to the left of the first significant digit. A decimal point should appear between the hundreds and the tens positions.

Error Listing Report

This report is to display all input records having a field or fields that failed to satisfy validation rules. Place an asterisk under each field in error. The format of this report is not fixed. Create your own. Make it neat and understandable.

New Employee and Termination Report (See Figure 2.)

This report is a combination of two reports. All hirings and terminations for the week are to be listed on this report. If the person is a new employee, all fields except Date of Termination and Cause of Termination are to be printed. If the person is leaving, all fields are to be printed.

1. Department (DEPT)
   A five-character, numeric field with zero suppression in the first two high-order positions.
2. Employee Number (EMP-NO)
   A six-character, numeric field with no zero suppression.
3. Employee Name (NAME)
   A twenty-character, alphabetic field.
4. Date of Hire
   A six-character, numeric field with month, day, year format. The month and day may have zero suppression in their high-order positions. Slashes (/) are to be inserted between the month and day and between the day and year.
5. Date of Termination
   This field has the same format as Date of Hire.
6. Type Employee (TYPE-EMP)
   This field is to identify the type of worker hourly or salaried. If an hourly worker, the letters HRLY are to appear in this field. If

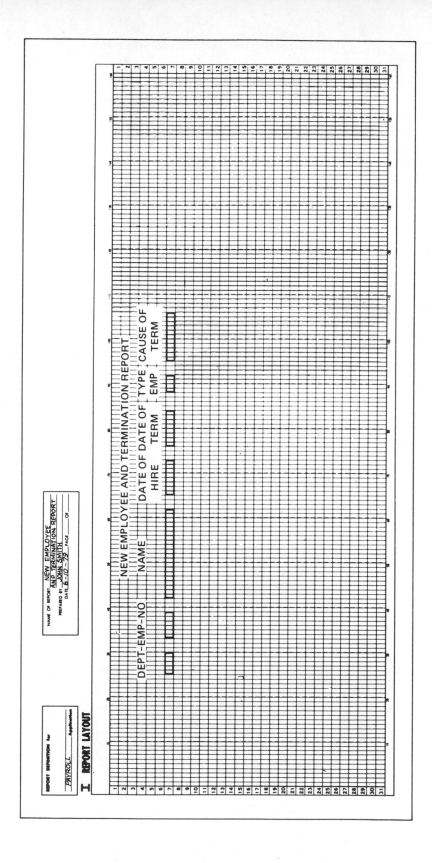

Figure 2. *Format of New Employee and Termination Report*

the employee is salaried, the letters SAL are to appear in this field.

7. Cause of Termination

There are three possible reasons for termination of pay:

   a. Transferred out of the realm of this payroll system. If an individual is transferred, the word TRANSFERRED is to appear in this column. (This is not a departmental transfer within the payroll system. The individual is being transferred out of this payroll system to another, yet within the same corporation.)

   b. Terminated by the employer. If the employer terminates the individual, the word TERMINATED is to appear in this column.

   c. Terminated voluntarily. If an employee leaves on his or her own accord, the word VOLUNTARY is to appear in this column.

Only one of these words can appear per line of print.

INPUTS TO THE SYSTEM

The inputs to the system are entered once each week. The formats of certain fields are identified and described below. These fields alone are not adequate to provide a complete solution to the problem. You must decide what other fields are needed to solve this payroll problem.

The following fields will always be present for a current employee:

1. Employee Number

   A six-character, numeric field.

2. Date of Reporting

   A six-character, numeric field with month, day, year format. This date cannot be greater than today's date or less than fourteen days prior to today. It is merely the date that the information is reported into the system, e.g., week-ending date.

3. Hours Worked for the Week

   A four-character, numeric field. This field cannot exceed the maximum number of hours in the week. The right-most position is tenths of hours. When used in computing, the field should be rounded to the nearest hour.

The following fields will always be present for a new employee:

1. Employee Number
    A six-character, numeric field.
2. Employee Name
    A twenty-character, alphanumeric field.
3. Birth Date
    A six-character, numeric field with month, day, year format.
4. Date of Hire
    A six-character, numeric field with month, day, year format. This date cannot exceed today's date or be less than fourteen days prior to today.
5. Hourly or Salaried Code
    A code that signifies whether an employee is hourly or salaried.
6. Rate or Wage
    A seven-character, numeric field. If the individual is a salaried employee, the field contains his or her weekly wage. The right-most two characters of the wage are cents positions. If the employee is on an hourly rate, the rate is in this field. The range is $2.25 to $15.00 per hour. The right-most two characters of the rate are cents positions. The employee cannot have more than one rate per week. Both wage and rate cannot appear in a given record.

REQUIREMENTS TO COMPLETE CASE: Complete or design all output and inputs to the system; create a data element dictionary; use decision tables as necessary; and develop a systems flowchart. Stay within the requirements of the problem.

Chapter 2

Samson
Manufacturing Inc.

Manufacturing companies are primarily involved with fabricating, machining, and/or assembling products to be sold to other manufacturers for further processing; to distribution companies for resale; and/or to retail outlets. To be profitable, a manufacturing company must manufacture the appropriate quantities of quality products in a timely and economical manner.

As with all cases in the book, this case is structured to be realistic and representative of the industry for which it is designed. The case is abbreviated so that it can be completed in a short period of time. In particular, the number of reports and terminal displays is reduced. So is the number of data elements. However, the case does include the most important and commonly recognized reporting in the manufacturing industry.

INTRODUCTION

You have been hired recently as a systems analyst by Samson Manufacturing Inc. The chief executive officers of Samson have decided to rework their production information systems. You have been assigned as the systems analyst to work on the project. Mr. Schroeder, President of Samson, has asked you to attend an executive meeting with the vice-presidents to discuss the new information system.

EXECUTIVE MEETING

At the executive meeting you were presented an abbreviated copy of Samson's organization chart (see Figure 2-1). During the meeting, Mr. Schroeder and Mr. Blair explained that Samson's production activities require information from two basic information systems:

- *Material Requirements Planning*—Provides the information necessary to manufacture the appropriate quantities and qualities of products in an economical and timely manner. Based upon the number and time when the final products need to be manufactured, the system should determine the material requirements and inform management as to what, how much, and when raw materials, parts, and subassemblies will be required.
- *Production Scheduling and Control*—Provides the information necessary to schedule and control the fabricating, machining, and assembling involved in the manufacturing process.

Mr. Schroeder and Mr. Blair further explained that due to increased volume and complexity in Samson's manufacturing activities, it is becoming increasingly important to have current, accurate information on material requirements and production scheduling and control. Therefore, Samson manufacturing wants

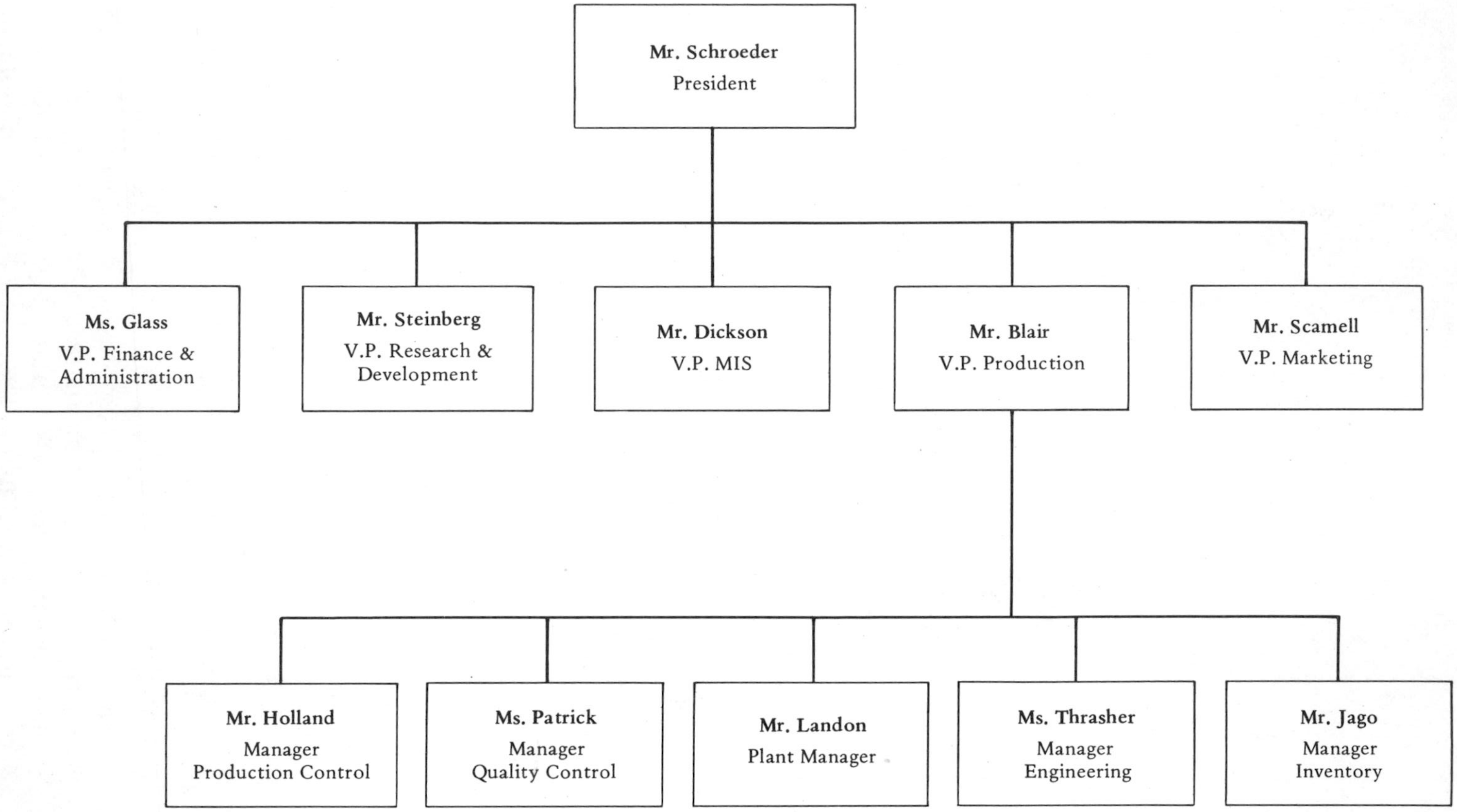

Figure 2-1. *Organization Chart for Samson Manufacturing*

to design a new production information system that incorporates online data entry as necessary.

You suggested that a project team be organized to conduct the necessary analysis and design the new information system. You requested that the project team be composed of key management personnel from the production departments affected by the new system.

Mr. Schroeder and Mr. Blair concurred with your suggestion. The following personnel were assigned to the project team:

Yourself, Systems Analyst and Project Leader
Mr. Landon, Plant Manager
Mr. Holland, Manager, Production Control
Ms. Patrick, Manager, Quality Control
Ms. Thrasher, Manager, Engineering
Mr. Jago, Manager, Inventory

Each team member was charged with determining the information requirements of his or her department. You were assigned overall responsibility for coordinating the team efforts and for developing the systems specifications.

FIRST PROJECT TEAM MEETING

You organized a meeting of the project team to initiate the project. You proposed that the team address the Material Requirements Planning (**MRP**) system first and then the Production Scheduling and Control (**PSC**) system. The project team agreed.

You explained that, for each system, the project team members needed to determine the reporting requirements of their departments. You explained that, once these have been determined, you will be able to "back into" the inputs and processing needed to generate the reports and terminal displays. You stressed that there is cost associated with generating computer-based information, and that only information necessary for operations and decision making should be included in the reports.

You offered to coordinate the work of the team members in determining reporting requirements.

SECOND PROJECT TEAM MEETING

After extensive analysis of the production information requirements at Samson, a second project team meeting was called. At this meeting, the project team consolidated the various information requirements for the MRP system into several reports. The reports are presented below.

Inventory Master List

Mr. Jago, Mr. Holland, Ms. Patrick, Ms. Thrasher, and Mr. Landon indicated that their respective departments need a weekly list of inventory items. Mr. Jago developed a sample report as a suggestion for the report format (see Figure 2-2).
      You asked Mr. Jago if PART NUMBER is always six digits. He said it is. Mr. Jago added that the PART DESCRIPTION can be up to fifteen characters long. He explained the remainder of the report as follows:

1. U/M—the unit of measure for an item. Options are:
    EA = Each   IN = Inch
    BX = Box   FT = Foot
    DZ = Dozen   YD = Yard
2. STOCK LOCATION—the warehouse location of an item. This field always contains two alphabetic characters followed by a digit.
3. ON HAND, ON ORDER, RESERVED, AVAILABLE, and REORDER POINT—numeric fields with a maximum value of 999,999. When printed, these fields are to have commas and zero suppression up to the least significant digit. RESERVED is the number of items reserved for scheduled production. AVAILABLE is computed by subtracting RESERVED from ON HAND. REORDER POINT is the level at which inventory should be ordered (i.e., when AVAILABLE is less than REORDER POINT, ordering should occur).
4. LEAD TIME—the number of working days (i.e., excluding weekends and holidays) required to replenish inventory.
5. PRO 1 PUR 2—a code indicating whether an item is produced by Samson or purchased from a vendor. Allowable code values are:
    1 = Produced by Samson
    2 = Purchased from vendor

DATE 5-07-80

INVENTORY MASTER LIST

PART NUMBER	PART DESCRIPTION	U/M	STOCK LOCATION	ON HAND	ON ORDER	RESERVED	AVAILABLE	LEAD TIME	REORDER POINT	PRO 1 PUR 2	DATE LAST ACTIVITY
106274	ARM ASSEMBLY	EA	AA1	5,000	10,000	2,000	3,000	15	8,000	1	5-01-80
106278	HUB	EA	AA3	2,000	0	0	2,000	10	1,000	2	4-05-80
107768	1/2 IN GEAR	EA	AA7	75,000	0	20,000	55,000	12	25,000	2	3-15-80
107774	.042 CRS COIL	EA	AB8	9,000	20,000	5,000	4,000	20	22,000	1	5-06-80
107852	COPPER TUBING	FT	BB1	125,000	0	35,000	95,000	10	35,000	2	2-05-80

Figure 2-2. Inventory Master List

49

6. DATE LAST ACTIVITY—the last date there was any activity (i.e., reorders, reductions, reservations). It is a calendar date formatted as follows:
MM-DD-YY

In conclusion, Mr. Jago mentioned that the report should be printed in part number order and include all parts that Samson stocks.

Extended Bill of Material

Mr. Landon, Mr. Holland, and Mr. Jago discussed the need for an extended bill of material report. They explained that an extended bill of material defines the parts and subassemblies required to manufacture a given assembly, subassembly, or end product. The extended bill of material is necessary for planning purposes—that is, to determine the components required for a given production order, the quantities required, and the lead time necessary for procuring or manufacturing each component.

Mr. Holland and Mr. Jago presented a sample report format for the extended bill of material (see Figure 2-3). Mr. Holland explained that there should be a separate page for each final assembly, subassembly, or end product. The part number and description of the assembly, subassembly, or end product should be listed under the report heading.

You asked for an explanation of the LEVEL reporting. Ms. Thrasher explained that a level number is a quantified representation used in engineering drawing. It defines the hierarchy of the various components that must be assembled to complete the product.

Mr. Jago explained that the LEVEL columns on one line may contain a value from 0 through 9. (The particular value is to line up with the corresponding digit in the LEVEL heading.) The value 0 is assigned to a completed assembly, subassembly, or end product that requires no further action. Parts and subassemblies that are subordinate to a level-0 component are assigned the value 1; parts and subassemblies subordinate to level-1 components are assigned the value 2; and so on. For example, in the sample extended bill of material report (see Figure 2-3), three level-3 items (644001, 687216, and 784234) are required to complete the level-2 INSERT (216871). The level-2 items 216871 and 467891 are required to complete the level-1 item HUB (106278). Four level-1 items are required to complete the assembly (level 0).

Second Project Team Meeting

DATE 5–15–80	EXTENDED BILL OF MATERIAL	PAGE 1

PART 106274 ARM ASSEMBLY

LEVEL 0 1 2 3 4 5 6 7 8 9	PART NUMBER	DESCRIPTION	U/M	QUANTITY REQUIRED	PRO CODE 1 PUR CODE 2	LEAD TIME
0	106274	ARM ASSEMBLY	EA	1.000	1	15
1	576843	STUD 1/2 IN	EA	5.000	2	2
1	432681	NUT 1/2 IN	EA	5.000	2	2
1	468721	ARM	EA	1.000	2	10
2	861231	.042 CRS COIL	EA	1.000	2	18
2	462111	.087 CDS NTRGN	IN	2.460	2	8
1	106228	HUB	EA	5.000	2	6
2	467891	SLEEVE 10 IN	EA	5.000	1	5
2	216871	INSERT	EA	10.000	2	4
3	784234	BEARING 1/4 IN	DZ	8.000	2	10
3	687216	HOUSING 1/4 IN	EA	10.000	1	5
3	644001	LUBRICANT	EA	1	2	5

Figure 2-3. *Extended Bill of Material*

  Mr. Holland indicated that all remaining information on the report except for QUANTITY REQUIRED is defined as discussed for the inventory master list. QUANTITY REQUIRED is the number of the U/M required for each part included in the extended bill of material.

Mr. Holland indicated that the extended bill of material should be printed in part number order and include all items.

Material/Parts Requisition

Mr. Landon and Mr. Jago reviewed the need for a material/ parts requisition. The requisition needs to be generated in response to a production order for a customer, or an internal production order for the manufacture of a part, assembly, subassembly, or end product. The material/parts requisition reporting controls material releases; lists all items included in a production order; determines quantity required using extended bill of material information; and assists in the timely movement of material by identifying where to obtain material, and where and when to deliver material.

Mr. Jago presented a sample report format for the material/ parts requisition (see Figure 2-4). Mr. Jago pointed out that much of the information included in the report was defined in previous reports. He defined the new information as follows:

1. CUSTOMER NUMBER—six-character field containing two alphabetic characters followed by four digits.
2. PRODUCTION ORDER—a five-digit number assigned by the computer when a production order is entered. It is used to uniquely identify a requisition.
3. SIGNED and DATE FILLED—headings to be filled in by the department that initiates a requisition when the materials are delivered.
4. FILLED BY—a heading to be signed by inventory personnel when an order is received, filled, and shipped to the requesting department.
5. DATE—the date the requisition is printed.
6. DELIVER TO—a four-character field. The first character may be A, B, C, D, E, or F. It is followed by a dash and two digits ranging from 01 to 55. This value is supplied by the requesting department manager when an order is placed.
7. DATE NEEDED—the date the material is needed. It is supplied by the requesting department manager when an order is placed.

Ms. Thrasher raised the question of how orders are to be placed. Mr. Landon explained that online terminals should be located in each department. The terminals should provide a display that helps department personnel to initiate orders.

SIGNED _DJM_	MATERIAL/PARTS	CUSTOMER NUMBER
DATE FILLED _5-15-80_	REQUISITION	AB1068
FILLED BY _FSG_		PRODUCTION ORDER
		10782

DATE	QUANTITY	PART NUMBER	PART DESCRIPTION	DATE NEEDED	PAGE
5–15–80	1000	106274	ARM ASSEMBLY	5–17–80	1

PART NUMBER	PART DESCRIPTION	STOCK LOCATION	DELIVER TO	U/M	QUANTITY
576843	STUD 1/2 IN	CC1	A–10	EA	5.000
432681	NUT 1/2 IN	DC2	A–10	EA	5.000
468721	ARM	BB7	A–10	EA	1.000
861231	.042 CRS COIL	AC5	A–10	EA	1.000

Figure 2-4. *Material/Parts Requisition*

You asked if requisitions need to be printed immediately after production orders are made. Mr. Landon indicated that they do. He asked if a hard-copy terminal could be located in the inventory department to facilitate this. You indicated it can. Mr. Jago said to be sure that all material requirements for a production order are reserved when the production order is processed.

Inventory Shortage Report

You asked Mr. Jago how he determines inventory shortages. He explained that he intends to review the inventory master list each week. He mentioned that it would be helpful if asterisks or a similar flag were used to highlight shortages on the report.

You asked Mr. Jago if reviewing the report once a week is timely enough, since orders are placed daily. Mr. Jago indicated that it is not, but he was reluctant to have a complete inventory master list printed daily.

You suggested that a daily exception report, listing only inventory items for which there are shortages, be printed. Mr. Jago thought this was an excellent idea. He suggested that the report have the same format as the inventory master list. You agreed to take care of it.

This concluded the second project meeting. You asked the team members to define the reporting requirements for the production scheduling and control (PSC) system during the next three weeks. Then the next project team meeting would be held.

THIRD PROJECT TEAM MEETING

After three weeks, the project team met to discuss reporting requirements for the PSC system. The reports are discussed below.

Production Orders

Mr. Landon and Mr. Holland first discussed the production order. They explained that the production order is the key transaction in production scheduling and control. It travels with an order throughout the production process and documents production operation standards.

Mr. Holland proposed a format for the production order (see Figure 2-5). He said much of the information on the report was defined in the MRP system. He defined the new information as follows:

1. QUANTITY ORDERED—the number of parts ordered. It can be up to 999999.
2. ISSUE DATE—the date a production order is generated by the computer system.
3. SCHEDULED START DATE, SCHEDULED COMPLETE DATE —the dates assigned by the production control department when a production order is entered through a terminal. The format of these fields is explained in 11 below.

PRODUCTION ORDER

PART NUMBER	DESCRIPTION	QUANTITY ORDERED	CUSTOMER NUMBER	ISSUE DATE	SCHEDULED START DATE	SCHEDULED COMPLETION DATE	PRODUCTION ORDER
106274	ARM ASSEMBLY	1000	AB1068	9-10-80	8-171	8-181	10782

OPERATION SEQUENCE	WORK CENTER	OPERATION CODE	SET-UP TIME	OPERATION STANDARD	EXTENDED OPERATION TIME	OPERATION DESCRIPTION	SCHEDULED FINISH
0001	A-10	BB-100	.75	.00120	1.95	RIVET 432614-781162	8-177
0002	A-10	BB-101	.50	.00150	2.00	BOLT 478111-681423	8-177
0003	A-11	CC-110	.25	.02160	21.85	MACHINE 786211-423881	8-179
0004	A-12	DD-567	.30	.00100	1.30	INSERT 644001	8-180
0005	A-12	DD-569	.05	.00160	2.10	INSERT 784234	8-181

Figure 2-5. *Production Order*

4. OPERATION SEQUENCE—the sequence in which production operations are to be performed. Sequence numbers are four digits, ranging from **0001**, in increments of 1, to the total number of operations required.

5. WORK CENTER—the physical location where work is performed. Work centers are coded as one alphabetic character, a dash, and two digits.

6. OPERATION CODE—the specific manufacturing process to be performed during a specific operation. The meaning of the code value can be looked up in a reference manual provided by engineering. The format of the code is two alphabetic characters followed by a hyphen and three digits.

7. SET-UP TIME—the time required to prepare a work center to perform an operation. It is expressed in hours or fraction of an hour. The largest set-up time is 15.00 hours.

8. OPERATION STANDARD—the standard time required to complete the operation for one unit. The time is expressed in hours, and may range from .00001 hour to **50.00000** hours. The time required is determined by time measurements conducted by the engineering department.

9. EXTENDED OPERATION TIME—the total standard time required to complete an operation. It is computed by multiplying the number of units to be produced by **OPERATION STANDARD** and adding it to **SET-UP TIME**. This figure may range from .01 to **800.00**.

10. OPERATION DESCRIPTION—a twenty-four character, alphanumeric field used to describe an operation.

11. SCHEDULED FINISH—the "manufacturing" days scheduled to complete a project. Manufacturing days (M-days) are the work days available during a year (i.e., they exclude weekends and holidays). The first M-day is 1. This number is incremented by 1 for each subsequent day until the end of the year. The format for the M-day is the last digit of the year (e.g., for 1983 the digit **3**) followed by a hyphen and the three digits of the M-day. The M-day calendar is set up by the production control manager.

Load Report

Mr. Holland indicated he needs a load report so that he can make scheduling decisions when production orders are entered (to set the start and complete dates). He said that he needs to know on a weekly basis the number of hours of work loaded into each work center, and the capacity of the work center. The report should project four weeks into the future.

Mr. Holland proposed a report format for the load report (see Figure 2-6). He explained the information on the report as follows:

1. CAPACITY—the total number of hours available at a work center during a given week. This figure is determined by the plant manager and the engineering department. It remains relatively stable.
2. LOADED—the total number of hours currently scheduled at a work center for a given week. It is computed by adding all production orders scheduled for a given week.
3. %—the loading level. It is computed by dividing LOADED by CAPACITY.
4. BEHIND SCH HRS—the amount by which LOADED exceeds CAPACITY during the next four-week period. It is computed by subtracting total LOADED from total CAPACITY. It is only printed when BEHIND SCH HRS is a positive number.
5. FUTURE LOAD—the sum of all production order requirements that are scheduled beyond the next four-week period.

Mr. Holland requested that the load report be printed in work center sequence.

Quantity Deviation Report

Ms. Patrick indicated the need for a quantity deviation report to report production deviations above or below acceptable ranges. She presented a sample report (see Figure 2-7) and explained the contents as follows:

1. PRODUCTION ORDER NBR—the standard number used for a production order.
2. PART NUMBER, OPERATION SEQUENCE, OPERATION CODE, and QUANTITY ORDERED—extracted from the production order under consideration.
3. PREVIOUS OPN QNTY—the number of items completed from the previous manufacturing operation (i.e., the number of items the previous operation started with, less any losses due to damage or shrinkage).
4. QUANTITY COMPLETED—the net quantity completed by the current work center (i.e., the number completed, less any losses due to damage or shrinkage).

DATE 6-1-80

LOAD REPORT

WORK CENTER	BEHIND SCH HRS	1ST WEEK			2ND WEEK			3RD WEEK			4TH WEEK			FUTURE LOAD
		CAPACITY	LOADED	%	CAPACITY	LOADED	%	CAPACITY	LOADED	%	CAPACITY	LOADED	%	
A-01	21.0	80	96.4	121	80	82.1	102	80	86.1	108	80	76.4	95	44.0
A-10		40	32.4	81	40	38.6	96	40	24.2	61	40	12.2	31	
A-11		160	140.5	88	160	155.0	103	160	72.6	45	160	108.7	67	
A-12		240	290.0	121	240	188.6	78	240	104.7	44	240	82.6	34	32.4
B-01		40	52.4	131	40	31.7	79	40	18.4	46	40	8.4	21	
B-12		320	100.0	31	320	152.6	47	320	91.6	29	320	77.5	24	128.9

Figure 2-6. *Load Report*

58

DATE 5-05-80				QUANTITY DEVIATION REPORT					
WORK CENTER	PRODUCTION ORDER NBR	PART NUMBER	OPERATION SEQUENCE	OPERATION CODE	QUANTITY ORDERED	PREVIOUS OPN QNTY	QUANTITY COMPLETED	%
A–10	10782	478120	0002	BB–101	1000	1000	975	97.5
A–11	10782	786220	0003	CC–110	1000	975	975	
A–12	10782	786221	0004	DD–567	1000	975	920	94.4

Figure 2-7. *Quantity Deviation Report*

5. %—the performance level as computed by dividing QUANTITY COMPLETED by PREVIOUS OPN QNTY. Not to be printed when % = 100.

Mr. Holland asked how the data for PREVIOUS OPN QNTY and QUANTITY COMPLETED are to be collected. You suggested that the production order be redesigned to capture these data at the work center as the production order is routed. Everyone agreed to this idea. You volunteered to redesign the production order to accommodate this change.

This was the last reporting requirement identified by the project team. You suggested that the project team finalize the formats of all reports and that you would present them to the executives of Samson. The project team members agreed. The meeting was terminated.

EXECUTIVE PRESENTATION

When the report formats were finalized, you presented the proposed reporting system at an executive meeting. Mr. Schroeder and the vice-presidents were impressed with the proposed system. They identified one additional report required by management.

Mr. Schroeder and Mr. Blair discussed the need for an effectiveness report. They explained that the report would facilitate analysis of the productivity of each work center and its supervision. They indicated the report should provide the following information about each work center on a weekly basis:

1. OPERATION CODE for each job performed during the week.
2. QUANTITY ORDERED for each job.
3. EXTENDED OPERATION TIME for QUANTITY ORDERED (including SET-UP TIME).
4. QUANTITY COMPLETED.
5. STANDARD OPERATION TIME EQUIVALENT. This is an adjusted figure to account for cases when QUANTITY COMPLETED is less than QUANTITY ORDERED. The standard operation time equivalent is computed by multiplying QUANTITY COMPLETED by OPERATION STANDARD (see Figure 2-5) and adding SET-UP TIME.
6. REPORTED OPERATION TIME required to complete the operation.
7. EFFECTIVENESS PERCENT, computed by dividing the STANDARD OPERATION TIME EQUIVALENT by the REPORTED OPERATION TIME.

Mr. Landon mentioned that there are currently no provisions for capturing reported time on operations. You suggested that the production order could be modified to allow this data to be captured manually. Everyone agreed to this approach.

You volunteered to design the effectiveness report. You asked what order the report should be in, and if all work centers should be reported. Mr. Blair indicated the report should be in work center order and should include all work centers. Mr. Schroeder indicated he does not need that much detail. He requested that his report include work center operations only under the following conditions. (HINT: This is a good application for a decision table.)

1. If an operation is less than 100 in quantity, only include it if the effectiveness percent is less than 75% or greater than 125% and/or if the quantity completed is less than 75% of the quantity ordered.
2. If an operation is between 100 and 1000 in quantity, only include it if the effectiveness percent is less than 85% or

greater than 115% and/or if the quantity completed is less than 85% of the quantity ordered.

3. If an operation is more than 1000 in quantity, only include it if the effectiveness percent is less than 95% or greater than 105% and/or if the quantity completed is less than 95% of the quantity ordered.

After discussing the new report, Mr. Schroeder directed you to complete the overall systems specifications for the new system. The meeting was then concluded.

REQUIREMENTS TO COMPLETE CASE

Complete or design all outputs from, and inputs to, the system; create a data element dictionary; construct any necessary decision tables; identify the necessary files; and flowchart the system.

In some instances, the exact format, input medium, and validation rules are not told here. You are at liberty to define such issues on a judgmental basis.

There are many ways this case can be expanded into a major time-consuming effort. To avoid this, stay within the requirements of the case.

Chapter 3

New National Bank

Banks are engaged primarily in providing services associated with checking accounts, savings accounts, and installment, commercial, and mortgage loans. The checking and savings accounts bring money into the bank from which the bank generates revenue by investing in loans. To be profitable, a bank's investment revenue must exceed its interest payments to savings accounts and its operating expenses (e.g., salaries, materials).

As with all cases in the book, this case is structured to be realistic and representative of the industry for which it is designed. The case is abbreviated so that it can be completed in a short period of time. In particular, the number of reports and terminal displays is reduced. So is the number of data elements. However, the case does include the most important and commonly recognized reporting in the banking industry.

INTRODUCTION

You have been hired recently as a systems analyst by New National Bank. The chief executive officers of New National have decided to develop a new information system to process DDA (demand deposit accounting, or checking accounts), savings accounts, and loans. You have been assigned as the systems analyst to participate in the project. Ms. Karpe, President of New National Bank, has asked you to attend an executive meeting with the vice-presidents to discuss the new information system.

EXECUTIVE MEETING

At the executive meeting you were presented a copy of New National's organization chart (see Figure 3-1). Ms. Karpe explained that Mr. Price, V.P. Operations, and his staff are primarily responsible for DDAs and savings accounts. Mr. Pender, V.P. Lending, and his staff are primarily responsible for making decisions on loans and administering loans. Mr. Adamson, Controller, and his staff take care of accounting and computer processing. Ms. Marr, V.P. Marketing, and her staff are responsible for researching and implementing programs to enhance the public image and market competitiveness of New National.

Ms. Marr explained that the research efforts of Mr. Morgan indicate a strong interest among existing and potential customers of New National for the bank to service customers by providing consolidated statements of banking activities. Under this approach, each customer would receive one statement reflecting the status of his or her DDAs, savings accounts, and loans each month.

Ms. Marr further pointed out that consolidating information about customers would also help the bank. A complete profile of a customer's various accounts would be useful for marketing promotions.

Mr. Pender indicated that consolidating information about customers would also help in making prompt decisions on loans.

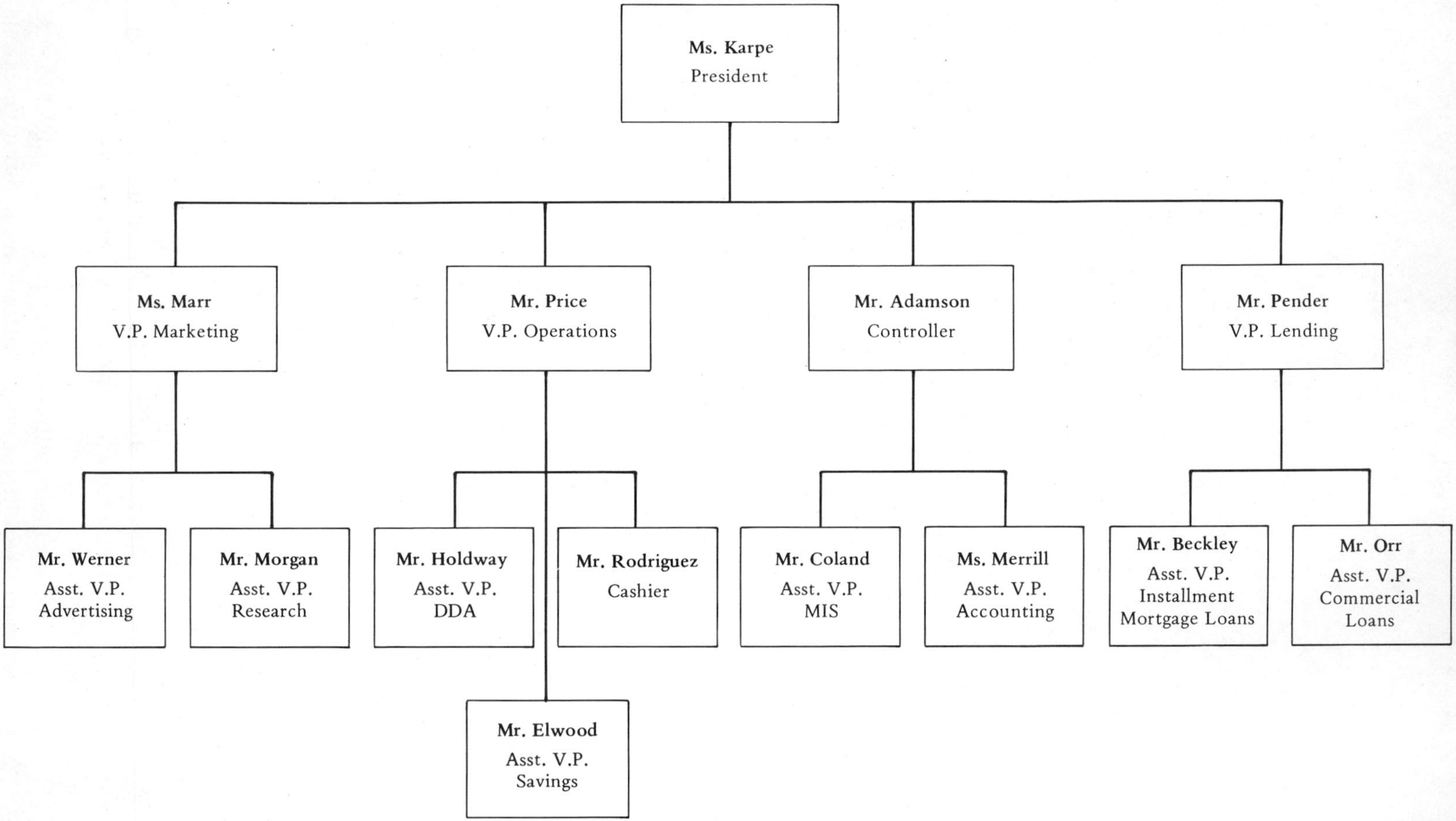

Figure 3-1. *Organization Chart for New National Bank*

Currently, the bank takes up to two days to review all the accounts of a customer (i.e., DDA, savings, and commercial, installment, and mortgage loans) before making a decision on a new loan. Since these are reported in different places, this activity is time-consuming, expensive, and results in poor service to the customer.

You asked what the current state of computer-based information systems is at New National. Mr. Adamson explained that New National is in the same position as many banks. It has all major applications on the computer. It uses magnetic–ink character recognition (MICR) equipment in processing checks and savings accounts.[1] All loan transactions are handled by prepunched cards given to customers when loans are originated. The customers then submit monthly payments with the corresponding computer cards.

Mr. Adamson indicated that the problem with the information systems at New National is that there is limited capability for integration of information. When DDA was computerized, the system was designed to meet the needs of the DDA department. When savings accounts were computerized, the system was designed to meet the needs of the savings accounts department. The DDA and savings accounts systems use different account numbers, so there is no convenient way to determine whether or not a customer who has a checking account with New National also has a savings account.

Subsequent design and implementation of systems for commercial, installment, and mortgage loans proceeded in the same manner as the DDA and savings accounts systems. Consequently, New National has customers with multiple checking, savings, and loan accounts, but it has no way to efficiently consolidate the information about each customer. One customer with multiple accounts looks like "multiple customers" to the bank. This results in redundant data capture, processing, storage, and reporting. New National often sends several statements (in different envelopes) to a customer.

Ms. Karpe explained that the inability to consolidate information for decision making and for the convenience of customers, and the obvious inefficiencies associated with processing redundancies, have prompted New National to undergo a major redesign of its banking applications to form an integrated information system. She further explained that she is totally supportive of using terminals for online processing when appropriate.

---

1. If you are not familiar with MICR technology, you are encouraged to visit a bank that uses MICR.

You suggested that a project team be organized to conduct the necessary analysis and to design the new information system. You requested that the project team be composed of high-level staff representatives from all major functions affected by the new system.

Ms. Karpe concurred with your suggestion and request. The following individuals were assigned to the project team:

Yourself, Systems Analyst and Project Leader
Mr. Holdway, Asst. V.P. DDA
Mr. Elwood, Asst. V.P. Savings
Mr. Beckley, Asst. V.P. Installment and Mortgage Loans
Mr. Orr, Asst. V.P. Commercial Loans
Mr. Rodriguez, Cashier
Ms. Merrill, Asst. V.P. Accounting

Each team member was charged with determining the processing and information requirements of his or her organizational area. You were assigned overall responsibility for coordinating the team efforts and for developing the systems specifications.

FIRST PROJECT TEAM MEETING

You organized a meeting of the project team to initiate the project. You asked each team member to determine the information requirements of his or her area. You explained that once these have been determined, you will be able to "back into" the inputs and processing needed to generate the reports. You also explained that there is cost associated with generating computer-based information and that only information necessary for operations and decision making should be included in the reports.

You indicated that you would coordinate the team efforts to insure integration of the information requirements. For example, basic customer demographic data (name, address, phone number, and so on) should be processed and stored *once* per customer, irrespective of the number of accounts the customer has.

SECOND PROJECT TEAM MEETING

After extensive analysis of the information requirements at New National, a second project team meeting was called. At this

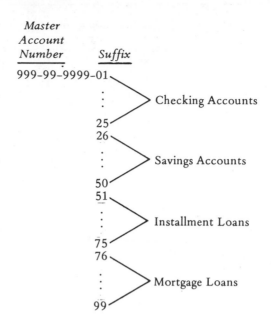

Figure 3-2. *Account Number Structure*

meeting, the project team consolidated the various information requirements into several reports. The reports are discussed below.

Consolidated Customer Statement

The first issue to be addressed was the concept of a consolidated customer statement. After considerable discussion, it was agreed that a uniform account structure is required to integrate the information from a customer's various accounts. The account structure agreed upon consists of a nine-digit master account number followed by a two-digit suffix (see Figure 3-2). Each customer will have one unique master account number. Suffixes will be used to uniquely identify different accounts of a particular customer.

After the account number structure was agreed upon, Mr. Holdway, Mr. Elwood, and Mr. Beckley presented their proposal for a consolidated customer statement (see Figure 3-3).

You asked for definitions of the contents of the consolidated statement. Mr. Holdway explained that the customer name and address are to be printed in the box so that, when the statement is folded, the customer name and address can be seen through an envelope window. He explained that the customer number is the master account number defined earlier. Since a statement may require more than one page, page numbers should be provided.

Mr. Holdway explained the remaining contents of the statement as follows:

NEW NATIONAL BANK
HOUSTON, TEXAS

MR. JOHN DOE
1218 BURNWOOD
HOUSTON, TEXAS, 77073

ACCOUNT NUMBER	BALANCE FORWARD	NUMBER OF CREDITS	TOTAL CREDITS	NUMBER OF DEBITS	TOTAL DEBITS	FEES AND EARNINGS	CLOSING BALANCE
Checking 01	432 75	2	225 60	4	375 00	2 00–	281 35
Checking 02	88 20	1	1,910 00	2	957 20	2 00–	1,039 00
Savings 26	1,670 00					8 35+	1,678 35
Inst. Ln. 41	580 30	1	125 10			5 80	461 00
Inst. Ln. 42	850 60	1	230 25			8 50	628 85
Comm. Ln. 61	7,725 30	2	1,605 60			64 37	6,184 07
Mort. Ln. 81	58,432 20	1	710 20			585 20	58,307 20

TRANSACTIONS

ACCOUNT NUMBER	AMOUNT	AMOUNT	AMOUNT	DATE
Checking 01	100 00 DP 10 15 CK 180 20 CK 777 00 CK	150 30 CK 164 35 CK	50 00 CK	06-10
Checking 02		1,910 00 DP	125 60 DP	06-21
Inst. Ln. 41	580 30 PM			06-05
Inst. Ln. 42	230 25 PM			06-18
Comm. Ln. 61	1,000 00 PM			06-28
Mort. Ln. 81	710 20 PM			06-27 06-28

CODE EXPLANATION—SEE OTHER SIDE

Figure 3-3. *Consolidated Customer Statement*

1. ACCOUNT NUMBER—the name of the category of the account followed by the account suffix. The account category is indicated by the value of the account suffix as follows:

    01–25 = CHECKING
    26–40 = SAVINGS
    41–60 = INST. LN. (INSTALLMENT LOAN)
    61–80 = COMM. LN. (COMMERCIAL LOAN)
    81–99 = MORT. LN. (MORTGAGE LOAN)

2. BALANCE FORWARD—the balance brought forward from the previous month. It should never exceed 9,999,999.

3. NUMBER OF CREDITS—the number of credit transactions during the month. It should never exceed 1000.

4. TOTAL CREDITS—the total credits for a month. It should never exceed 9,999,999.

5. NUMBER OF DEBITS—the number of debit transactions during the month. It should never exceed 1000.

6. TOTAL DEBITS—the total debits for a month. It should never exceed 9,999,999.

7. FEES AND EARNINGS—the expenses and/or revenue of an account during a month. Possible FEES and EARNINGS follow:

    CHECKING—$2.00 service fee unless customer has $200 or more in checking, in which case there is no fee. The customer pays an additional $2.00 fee if more than twenty transactions (i.e., debits and/or credits) occur on a checking account in a month. The additional fee does not apply to accounts with $200 or more in checking. Fees are always followed by a minus sign.

    SAVINGS—A savings account earns the interest rate assigned to the customer's account. Interest earnings are prorated on a monthly basis. The customer is charged a $2.00 fee if a savings account has more than five transactions in a month. Earnings are always followed by a plus sign.

    INST. LN., COMM. LN., MORT. LN.—A fee—the interest charge—is prorated on a monthly basis and is always followed by a minus sign.

8. CLOSING BALANCE—Computed by adding TOTAL CREDITS and EARNINGS to; and subtracting TOTAL DEBITS and FEES from, BALANCE FORWARD.

9. TRANSACTIONS—itemized details for each transaction: AC-COUNT NUMBER, AMOUNT, and DATE (DATE is formatted mm-dd). The form allows up to four transactions to be listed

on the same line. If more than four transactions occur on one day, the next line is used to display additional transaction(s). Each transaction AMOUNT is followed by a transaction code. The possible code values and their meanings are:

DP = Deposit to a savings or checking account
CK = Check written against a checking account
WD = Withdrawal from a savings account
PM = Payment on a loan
LN = Additional principal added to an existing loan
        or initial principal for a new loan

The project team members were satisfied with the format of the consolidated customer statement. It was approved.

MICR-Entry Report

Mr. Rodriguez discussed the need for an MICR-entry report. He explained that transactions are sent in batches to the data processing department. He said that, for control purposes, each batch of transactions has a header record that indicates the number of transactions and the total dollar value for the batch. As each batch of transactions is run through the MICR sorter and entered into the computer, the cashier's office needs a list of all items and control totals to check against the batch totals.

He proposed a format for the MICR-entry report (see Figure 3-4). Since the contents of the report are straightforward, few questions were asked. Mr. Rodriguez did explain that the transactions are entered into the computer on the first pass of the sorter. Therefore, the transactions will not be sorted until after the MICR-entry report has been generated. The transactions are to be listed sequentially, "three-up" across the page. The ACTUAL TOTAL and ACTUAL ITEM COUNT are computed as the transactions are processed by the computer. They are compared to the HEADER TOTAL and HEADER ITEM COUNT to compute each DIFFERENCE.

Posting Journal

Mr. Holdway indicated that he, as well as the other managers, needs a daily posting journal that reflects all transactions performed against customer accounts. He proposed a format for the report (see Figure 3-5). The posting journal should list all customer

MICR-ENTRY REPORT					
BATCH NO. 102					
ACCOUNT NO.	AMOUNT	ACCOUNT NO.	AMOUNT	ACCOUNT NO.	AMOUNT
586–24–7123–01	8.93	782–42–4687–01	1,689.40	784–36–4777–01	78.40
478–96–8742–01	486.47	432–68–7844–20	50.00	742–78–4928–05	32.00
681–42–7842–03	620.00	333–78–4932–21	872.00	784–36–4777–01	781.20

HEADER TOTAL	6,889.80	HEADER ITEM COUNT	48
ACTUAL TOTAL	6,889.80	ACTUAL ITEM COUNT	48
DIFFERENCE	.00	DIFFERENCE	0

Figure 3-4. *MICR-Entry Report*

accounts that had activity the previous day. He said most of the report contents were already defined in other reports. He defined new information (and previously defined information presented in a different form) as follows:

1. DATE—the date of the report.
2. ACCT—the classification and the number of the account. To save space, this information is abbreviated as follows:

   CK = Checking account
   SA = Savings account
   IL = Installment loan
   CL = Commercial loan
   ML = Mortgage loan

3. OLD BALANCE—the customer balance prior to transaction posting. In the case of checking and savings accounts, the balance is a customer asset. In the case of loans, the balance is the principal remaining on the loan.
4. DATE LAST TRAN—the last date there was any transaction activity for the account. It is updated whenever a transaction occurs.

POSTING JOURNAL

DATE 7-05-80

CUSTOMER	ACCT	OLD BALANCE	DATE LAST TRAN	AMOUNT	AMOUNT	AMOUNT	AMOUNT	AMOUNT	NEW BALANCE
772-43-6871	CK01	362.30	6-10-80	32.00CR	28.00CR	10.00DB			412.30
BILL GLASS									
781-24-6371	CK01	1,781.20	06-01-80	100.00CR	22.00DB	20.00CR	30.00CR	2,000.00CR	
CHERYL GLASS				1,000.00DB	50.00CR				2,959.20
532-46-8711	SA26	5,742.80	05-05-80	1,000.00CR	18.00DB				6,742.80
LINDA GLASS	IL42	1,642.80	06-05-80	42.50CR					1,600.30
711-41-6111	ML81	38,572.31	06-05-80	420.00CR					38,152.31
WANDA GLASS	CL62	10,784.70	06-06-80	500.00CR	30.00CR	78.00CR	100.00CR	50.00CR	9,186.70
				720.00CR	28.00CR	92.00CR			

Figure 3-5 Posting Journal

5. AMOUNT—a detail listing of all debits and credits to the account. The symbol DB (for debit) or CR (for credit) follows each AMOUNT value.
6. NEW BALANCE—the new balance for the account value after all transactions have been posted. It is computed by adding all credits to, and subtracting all debits from, the OLD BALANCE.

The posting journal is to be printed twice: once in customer number order, and once in alphabetical order by customer name. Note that under the heading CUSTOMER, the customer number is printed first and the corresponding name is printed right below it.

Trial Balance

Next Ms. Merrill indicated that she needs trial balance figures after all transactions have been posted. She requested a trial balance report, listing all categories of accounts (i.e., DDA, savings, installment loans, commercial loans, and mortgage loans). It should show the balance for each account within a category (i.e., DDA01, DDA02, . . . , DDA25).

You asked if the balance should include the balances for all accounts or just those experiencing transactions the previous day. She said it should be the balance of all accounts. For example:

	DDA01	$2,432,628.03
	02	897,432.20
		1,785,672.30
		—
		—
		—
	25	786,432.30
TOTAL		$32,468,721.60

You indicated that the trial balance could be added to the posting journal report. The program could sequentially pass all accounts, posting to accounts having transactions, and picking up balances for all accounts. The balances could then be printed on the last page or two of the posting journal.

Everyone agreed that this would be an efficient way to accomplish reporting for the posting journal and the trial balance. You offered to design a format for the trial balance portion of the combined report.

This concluded the second project team meeting. The project team agreed to meet in three weeks to continue defining reporting requirements.

THIRD PROJECT TEAM MEETING

Customer Master Display

At the third project meeting, the first report discussed was a customer master terminal display. All project team members had expressed a need for a computer terminal display whereby accurate information about a given customer could be obtained upon request. After considerable discussion, the project team agreed on a format (see Figure 3-6). Much of the customer master information was defined in previous reports. The new information was defined as follows:

1. H-PHONE, 0-PHONE—home and office phone numbers, respectively.
2. HOME—the current housing of the customer. Possible values are: 1 = Owns; 2 = Rents.
3. SEX—coded as follows:
    1 = Male
    2 = Female
    3 = Couple
4. MARITAL—coded as follows:
    1 = Single
    2 = Married
    3 = Divorced
    4 = Widowed
5. CHILD—the number of children.
6. EMPLOYER—the name of the employer of the first name given in the customer name (in Figure 3-6, of Joe).
7. EMPLOYER-S—for married customers, the employer of the spouse (in Figure 3-6, of Devra).
8. YRS-EMP, YRS-EMP-S—the number of years employed by current employer (S designates spouse and is only used when applicable).
9. INCOME, INCOME-S—the income (S designates spouse and is only used when applicable).
10. DDA—checking account numbers.
11. BALANCE—the current balance in a checking account.

CUSTOMER MASTER

CUSTOMER NO. 585-12-7864

NAME AND ADDRESS		HOME	MARITAL	EMPLOYER
MR. JOE AND MS. DEVRA CHILDS		1	2	PIZZA HUT
1218 MAIN	H-PHONE 712-6841			
HOUSTON, TEXAS 77073	O-PHONE 744-3177	SEX	CHILD	EMPLOYER-S
		3	2	EXXON

DDA	BALANCE	YR	NSF	SA	BALANCE	YR	INT-R	YRS-EMP	
01	432.20	77	0					3	
02	25.10	78	1	26	1,562.50	78	.065	YRS-EMP-S	3

INCOME 16,000
INCOME-S 16,000

LOANS	O-BALANCE	C-BALANCE	PAYMENT	INT-RATE	INT-PD	PMTS	LT-PMTS	DT-LAST-PMT	PMT-DUE
41	785.00	582.75	35.00	.120	85.62	12	0	05-03-80	06-05-80
42	2,200.00	1,871.50	136.00	.120	141.70	18	1	05-03-80	06-05-80
81	39,500.00	38,500.00	398.00	.095	2,658.30	360	0	05-03-80	06-05-80

Figure 3-6. Customer Master Display

12. YR—the year an account was opened.
13. NSF—the number of checks written when there were not sufficient funds.
14. SA—savings account numbers.
15. INT-R—the interest rate (entered through terminal by the savings department when an account is opened).
16. LOANS—loan numbers.
17. O-BALANCE, C-BALANCE (LOAN)—O is the original principal borrowed; C is the principal remaining.
18. PAYMENT—the monthly payment (entered by a loan officer when loan is originated).
19. INT-RATE—the interest rate on a loan (entered through terminal by a loan officer when loan is originated).
20. INT-PD—the interest paid on the loan since loan origination (computed after each payment).
21. PMTS—the number of payments that have been made.
22. LT-PMTS—the number of payments that have been late (fifteen days later than the due date).
23. DT-LAST-PMT—the date the last payment was made.
24. PMT-DUE—the date the next payment is due (advanced one month each time a payment is made).

The project team members want to be able to access customer master information by either customer number or customer name.

Control Reports

Mr. Holdway explained that he needs a daily report indicating all checking accounts actioned without sufficient funds (NSF accounts). He indicated that if a customer's debit(s) exceeds his or her balance, the debit(s) is not to be posted. An exception report listing customers with NSF accounts is required.

You indicated that you were glad he mentioned this requirement. It means that all transactions have to be sorted so that credit transactions are posted before debit transactions. Otherwise, a customer may be charged erroneously for an NSF account. Mr. Holdway agreed.

You proposed that the exception report be prepared as a by-product of creating the posting journal. You asked Mr. Holdway what the report should contain. He requested that it contain customer number and name, address, office and home phone numbers, old balance, and the transactions posted. You agreed to design a report format.

Next Mr. Beckley and Mr. Orr indicated they need a monthly exception report of all customers whose loans are more than thirty days delinquent in payments. You indicated that since date-of-last-payment and next-payment-due-date are already available, a simple comparison can be used to detect delinquent accounts. A program could be run at the end of the month to check the dates and report all exceptions.

Mr. Beckley and Mr. Orr agreed. They asked if installment, commercial, and mortgage loan categories could be separated on the report. You said this could be accomplished by sorting the data before printing. You asked what information the report should contain. They indicated they need customer number and name, address, office and home phone numbers, current balance, date-of-last-payment, and next-payment-due-date. You agreed to design a report format.

This was the final reporting requirement identified by the project team. You suggested that the project team finalize the formats of all reports and that you would present them to the executives of New National. The project team agreed. The meeting was concluded.

EXECUTIVE PRESENTATION

When the report formats were finalized, you presented the proposed reporting system at an executive meeting. Ms. Karpe and the vice-presidents were impressed with the proposed system. They identified one special management report they need to monitor special activity. Ms. Karpe indicated that she and the vice-presidents want a daily report, listing customers in the following categories. (HINT: This is a good application for a decision table.)

1. Any customer whose checking account balance is $20,000 or more.
2. Any customer whose checking account balance is $10,000 or more who does not have a savings account.
3. Any customer who wrote a check for $10,000 or more.
4. Any customer who had $20,000 or more in savings during the day.

5. Any customer who deposited more than $30,000 in savings or withdrew more than $20,000.
6. Any customer with more than $10,000 in savings who does not own a home.
7. Any customer who has an installment loan with more than $10,000 balance.
8. Any customer who has a commercial loan with more than $500,000 balance.
9. Any customer who has a mortgage loan with more than $150,000 balance.

Ms. Karpe explained that the bank wants to know or become familiar with customers in these categories. She explained that the customer master display format would be satisfactory for the report. She wants the report to be printed in alphabetical order by customer name. She said that a coding system should be set up to explain why a customer is listed on the report. For example, a code value of 1 following the customer information could mean a customer with over $100,000 in savings. Obviously, more than one code value may be applicable to a customer at any point in time. You agreed to design such a coding system.

After this discussion, Ms. Karpe directed you to complete the overall systems specifications for the new system. The meeting was then concluded.

REQUIREMENTS TO COMPLETE CASE

Complete or design all outputs from, and inputs to, the system; create a data element dictionary; construct any necessary decision tables; identify the necessary files; and flowchart the system.

In some instances, the exact format and validation rules are not told here. You are at liberty to define such issues on a judgmental basis.

There are many ways this case can be expanded into a major time-consuming effort. To avoid this, stay within the requirements of the case.

Chapter 4

Memorial Hospital

Hospitals are called upon to provide an increasing variety of patient services. Medical, surgical, chemical, biological, therapeutic, dietary, housekeeping, and transportation techniques are exemplary of the complex array of services offered. To be effective, hospitals are having to increase the speed and personalization with which they respond to demands for hospital resources.

As with all cases in this book, this case is structured to be realistic and representative of the industry for which it is designed. The case is abbreviated so that it can be completed in a short period of time. In particular, the number of reports and terminal displays is reduced. So is the number of data elements. However, the case does include commonly recognized hospital reporting.

INTRODUCTION

You have been hired recently as a systems analyst by Memorial Hospital. The chief administrators of Memorial have decided to develop a new information system for the hospital. You have been assigned as the systems analyst to participate in the project. Mr. Dock, Hospital Administrator, and Dr. Cornette, Chief of Staff, have asked you to attend an executive meeting of the hospital to discuss the new information system.

EXECUTIVE MEETING

At the executive meeting you were presented an abbreviated version of Memorial's organization chart (see Figure 4-1). During the meeting it was explained that the hospital is divided into two major organizational groups. The physicians, headed by Dr. Cornette, are ultimately responsible for their patients. The group headed by Mr. Dock provides the clinical, nursing, and administrative support required by the physicians to service their patients.

Mr. Dock explained that operational management of the many services needed for a large population of patients requires coordinated reporting from Memorial's information systems. Also, individual services must be accounted for in detail. Each patient is entitled to an explanation of each charge he or she incurs. Prompt financial action from third-party coverage (e.g., insurance companies) is dependent on detailed statements showing the proper division of costs between patients and third parties.

Dr. Cornette explained that Memorial needs to consider not only the schedules and inventory replenishments that are a part of 24-hour operation but also the reallocation of existing resources and plans for those that are not yet available. Accordingly, the hospital's information system must tally the frequency of demand for various services and print cost summaries for departments providing services.

Mr. Dock added that financing the long-term capital re-

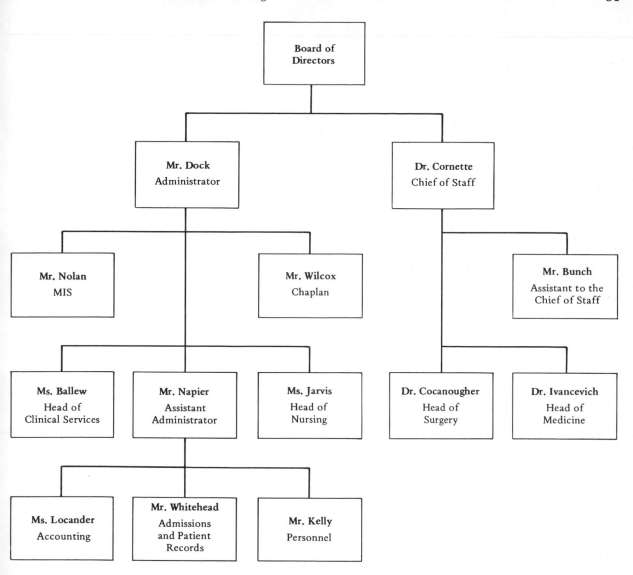

Figure 4-1. *Organization Chart for Memorial Hospital*

quired for hospital expansion and improvement is most success-
fully approached when the community is supportive of the hospital's efforts. An understanding community attitude is fostered by
regular and specialized explanations of Memorial's contribution to
the community. Statistical summaries of hospital services provided can play a key role in keeping the community informed.

Mr. Dock and Dr. Cornette went on to explain that they
want a new information system that addresses hospital and patient

reporting. They indicated that, due to the critical nature of the accuracy and timeliness of information, they favor using online terminals as appropriate.

You suggested that a project team be organized to conduct the necessary analysis and design the new information system. You requested that the project team be composed of high-level staff representatives from all major hospital functions affected by the new system.

Both Mr. Dock and Dr. Cornette concurred with your suggestion. The following individuals were assigned to the project team:

Yourself, Systems Analyst and Project Leader
Mr. Bunch, Assistant to the Chief of Staff
Ms. Ballew, Head of Clinical Services
Mr. Napier, Assistant Administrator
Ms. Jarvis, Head of Nursing

Each team member was charged with determining the information requirements of his or her organizational area. You were assigned overall responsibility for coordinating the team efforts and for developing the systems specifications.

FIRST PROJECT TEAM MEETING

You organized a meeting of the project team to initiate the project. You asked each team member to determine the reporting requirements of his or her area of the organization. You explained that once these have been determined, you will be able to "back into" the inputs and processing needed to generate the reports. You also explained that there is cost associated with generating computer-based information and that only information necessary for operations and decision making should be included in the reports.

You offered to coordinate the work of team members to insure integration of the information reporting requirements. For example, patient data are required by the responsible physician, various clinics, Admissions, Accounting, Nursing, and so on. Accordingly, each area needs to have convenient access to patient data without inefficient duplication of record keeping.

SECOND PROJECT TEAM MEETING

After extensive analysis of the information requirements at Memorial Hospital, a second project team meeting was called. At this meeting, the project team began consolidating the various information requirements into several reports. The reports are presented below.

Daily and Monthly Revenue Report

Mr. Napier discussed the need for a daily revenue report of all dollar transactions. He provided a sample report as a suggestion for the report format (see Figure 4-2).

You asked Mr. Napier whether PATIENT NUMBER is assigned by the hospital and whether it is always five digits long. He

Figure 4-2. *Daily Revenue Report*

					MEMORIAL HOSPITAL		
DATE 5–31–80					DAILY REVENUE REPORT		PAGE 1
PATIENT NUMBER	LOCATION	FINANCIAL STATUS	COST CENTER		ITEM DESCRIPTION	AMOUNT	TOTAL
54321	3021–1	BLUE–CR	112		ROOM–PRIVATE	80.00	
	3021–1	BLUE–CR	125		PHARMACY–SOLUTIONS	15.00	
	3021–1	SELF–PAY	112		TELEVISION	4.00	
	3021–1	BLUE–CR	132		X-RAY–CHEST	16.00	
							115.00
71623	2125–2	MEDICARE	112		ROOM–4 RM SUITE	40.00	
	2125–2	MEDICARE	101		PATHOLOGY	26.50	
	2125–2	MEDICARE	025		PHARMACY–PENICILIN	14.80	
							81.30

TOTAL PATIENTS	TOTAL CHARGES
620	$36,732.50

indicated that **PATIENT NUMBER** is assigned by the Admissions Office and that it is always five digits long.

Mr. Napier explained the contents of the remainder of the report as follows:

1. DATE—the day for which the report is printed. Its format is mm-dd-yy.
2. PAGE—the page number of the report. It is initialized to 1 and incremented by one for each page.
3. LOCATION—a five-digit field (with a hyphen preceding the last digit) indicating the room and bed a patient is assigned to. The first digit indicates the floor where the room is (there are five floors). The next three digits indicate the room (there are up to 125 rooms on each floor). The last digit indicates which bed in the room a patient is in.
4. FINANCIAL STATUS—the source of payment for hospital services. This information is coded on patient records to save space. Therefore, table look-ups are to be used in printing the report. The possible code values and their meanings are:

   1 = BLUE CROSS
   2 = OTHER-BC (a company other than Blue Cross)
   3 = MEDICARE
   4 = MEDICAID
   5 = COMM-INS (commercial insurance)
   6 = WORK-COMP (workmen's compensation)
   7 = WELFARE
   8 = SELF-PAY
   9 = OTHER

5. COST CENTER—a three-digit code that uniquely identifies a cost center (pharmacy, x-ray, rooms, etc.). The hospital currently has 75 cost centers, identified by the code values from 101 to 175. The name of a cost center can be looked up in a list provided by the Assistant Administrator. (Only the code values are required on this report.)
6. ITEM DESCRIPTION—the description of the service provided by the cost center. It can be up to twenty-five characters long.
7. AMOUNT—the dollar charge for an item.
8. TOTAL—the total day's charges for a patient. It is computed by adding the **AMOUNT** values for the patient.
9. TOTAL PATIENTS—the total number of patients having transactions during the day. **TOTAL PATIENTS** is printed on the last page of the report.

10. TOTAL CHARGES—the total of all charges during the day (i.e., the sum of the TOTAL column). It is printed on the last page of the report.

You asked Mr. Napier where ITEM DESCRIPTIONS and AMOUNTS come from. He explained that, as patients receive services during the day, the cost centers fill out forms indicating patient names and numbers, cost center codes, and services rendered. These forms go to the Accounting Department during the day. A charge is assigned for each service in accordance with the hospital's pricing manual.

You suggested that the transactions be keypunched at the end of the day and processed overnight. This approach was agreeable to the project team.

Mr. Napier added that the hospital needs a monthly version of the daily revenue report. You indicated this would be no problem. The daily transactions can be stored and a report generated at the end of the month. (Note that this second report requires a different heading.) Mr. Napier indicated that both reports should be in patient number sequence.

Room Control Report

Ms. Jarvis and Mr. Napier discussed the need for a daily room control report. They explained that such a report would facilitate bed planning and staffing and provide the statistical basis for facility and room expansion. Mr. Napier pointed out that the Admissions Office desperately needs accurate information on admissions, discharges, and transfers in order to control room assignments.

Ms. Jarvis and Mr. Napier provided a sample format for the room control report (see Figure 4-3). Some of the contents of the report were defined in the previous report. They defined the new items as follows:

1. ACCOM—the room accommodations. Possible code values and their meanings are:

    PR = PRIVATE
    SP = SEMI-PRIVATE
    3R = 3-BED WARD
    4R = 4-BED WARD
     IC = INTENSIVE CARE

DATE 05-31-80		ROOM CONTROL REPORT					PAGE 1
ROOM NUMBER	ACCOM	PATIENT NAME	NUMBER	SEX	AGE	ADMIT DATE	EXPECTED DISCHARGE
1001–1	PR	JOHNSON, GEORGE	58723	M	38Y	05–28–80	06–01–80
1002–1	PR	LITZSINGER, JUDY	46115	F	31Y	05–20–80	06–05–80
1003–1	SP	FRANCISCO, AL	78711	M	22Y	05–31–80	
1004–2	SP	IDOL, CHARLES	43244	M	28Y	05–29–80	06–10–80

Figure 4-3. *Room Control Report*

2. PATIENT NAME—up to thirty characters, printed as follows:

   LAST NAME, FIRST NAME, MIDDLE NAME OR INITIAL

3. SEX—coded as follows:

   M = MALE
   F = FEMALE

4. AGE—the patient's age. When the age is followed by Y, the age is in years. When the age is followed by M, the age is in months.
5. ADMIT DATE—the date the patient was last admitted.
6. EXPECTED DISCHARGE—the date the patient is expected to be discharged. If this is not known, the field is blank.

   The report is to be printed in room number order.

Scheduled Admissions Report

Mr. Napier indicated that the Admissions Office needs a daily scheduled admissions report so that office personnel can prepare for admissions. Mr. Napier provided a sample report format (see Figure 4-4). Much of the information contained in the report had been defined. Mr. Napier defined the new items as follows:

1. DOCTOR—the doctor assigned to a patient. This is supplied to the Admissions Office when a reservation is made.
2. EXPC DATE—the date a patient is expected to be admitted. The format is mm-dd.
3. TELEPHONE—the phone number where the patient can be reached. It consists of the area code followed by the phone number.
4. THIRD PARTY—the primary third party responsible for financial payment. It can be up to twelve characters long.

Mr. Napier indicated the report is to be sequenced by PATIENT NAME (alphabetically) within EXPC DATE.

Figure 4-4. *Scheduled Admissions Report*

		MEMORIAL HOSPITAL							
DATE 05-31-80		**SCHEDULED ADMISSIONS REPORT**							**PAGE 1**
PATIENT NAME	PATIENT NUMBER	DOCTOR	EXPC DATE	ACCOM	AGE	SEX	TELEPHONE	THIRD PARTY	
ADAMSON, RAY	46781	WERNER, M.A.	06-01	PR	16Y	M	713-461-7821	BLUE CROSS	
MARR, NANCY	37842	PENDER, R.P.	06-01	SP	32Y	F	713-444-3681	AETNA	
PRICE, JOHN H.	17784	CHILDS, J.	06-01	PR	11M	M	612-336-8791	NCR GROUP	
WILCOX, HARRY	33614	PENDER, R.P.	05-01	4R	45Y	M	713-456-8217	MEDICARE	

Physician-Patient Report

Mr. Bunch discussed the need for a daily physician-patient report to insure that all physicians have an accurate account of all their patients currently in the hospital. He provided a sample format for the report (see Figure 4-5).

Mr. Bunch pointed out that much of the report contents were defined in previous reports. The new information was defined as follows:

1. EXT—the hospital phone extension of a patient. It is four digits long and always begins with a 3.
2. STREET ADDRESS—the home street address of a patient. It can be up to fifteen characters long.
3. CITY-STATE-ZIP—the home city, state, and zip code of a patient. It can be up to thirty characters long.

Mr. Bunch requested that the report be in LOCATION sequence to assist physicians in making their daily rounds.

Figure 4-5. *Physician-Patient Report*

MEMORIAL HOSPITAL PHYSICIAN—PATIENT REPORT				
DATE 05-31-80		PHYSICIAN—WERNER, M.A.		PAGE 1
LOCATION	EXT	PATIENT NAME	STREET ADDRESS	CITY—STATE—ZIP
1025-2	3162	ESTES, NAOMI	1146 PONDEROSA	HOUSTON, TEXAS 77116
1026-1	3175	JENKINS, MILT	1146 OLD OAKS	HOUSTON, TEXAS 71384
1030-3	3160	WILLIAMSON, MARY	1816 ATASCOCIDA	PHOENIX, ARIZONA 16871
1040-1	3180	DEERE, PEGGY	1106 POCATELLO	HOUSTON, TEXAS 78113

After Mr. Bunch's request, the project team meeting was concluded. A third project team meeting was scheduled to discuss patient records information and billing.

THIRD PROJECT TEAM MEETING

Patient Master Display

At the third project team meeting, the first issue discussed was a terminal display to be used for input and retrieval of patient information. After much deliberation, the project team agreed that the information below should be contained in the terminal display. (The information that is not contained in previous displays or is not self-explanatory is defined.)

1. PATIENT NUMBER
2. PATIENT NAME
3. STREET ADDRESS
4. CITY-STATE-ZIP
5. TELEPHONE
6. RESPONSIBLE PARTY—a person to be contacted in case of an emergency. This is a twenty-five character field; last name precedes first name.
7. RESPONSIBLE PARTY'S STREET ADDRESS
8. RESPONSIBLE PARTY'S CITY-STATE-ZIP
9. RESPONSIBLE PARTY'S TELEPHONE
10. SOCIAL SECURITY NUMBER
11. SEX
12. MARITAL STATUS
13. BLOOD TYPE—one of the following values:

    A+ or A−
    B+ or B−
    AB+ or AB−
    O+ or O−

14. BIRTH DATE—formatted mm-dd-yy.
15. ALLERGY CODES—two-digit code values, separated by commas. There can be up to five code values.
16. RELIGION CODE—a one-digit code value, ranging from 1 to 5. Values are defined as follows:

1 = Protestant
2 = Catholic
3 = Jewish
4 = Other
5 = Not applicable

17. LOCATION—used only when patient is in the hospital.
18. EXPECTED DISCHARGE—used only when patient is in the hospital.
19. EXPC DATE
20. ADMIT DATE
21. DOCTOR
22. DISCHARGE DATE—the date a patient was last discharged. It is used only when applicable.
23. THIRD PARTY
24. FINANCIAL STATUS
25. EXT
26. ADMIT DIAGNOSIS DESCRIPTION—a twenty-character field for documenting doctor's preliminary diagnosis.
27. ADMIT DIAGNOSIS CODE—a standardized six-digit code that can be cross-referenced to a diagnosis code book.
28. FINAL DIAGNOSIS DESCRIPTION—a twenty-character field for documenting doctor's final diagnosis.
29. FINAL DIAGNOSIS CODE—same as 27.
30. FINAL DIAGNOSIS DATE
31. DATE OF DEATH—used only when applicable.

The project team asked you to design formats for the patient master display. You pointed out that, depending on the terminal screen size and the format used, two or more "pages" (screen layouts) may be required to display all of the information. This was acceptable to the project team.

You asked whether patient information exists for individuals not currently in the hospital. Mr. Napier indicated it does. Any individual admitted to the hospital becomes a permanent part of the hospital's records. You suggested it may be wise to move records that have been inactive (i.e., records for patients who have not been re-admitted for a prolonged period of time) from the online disk file to less expensive, long-term storage such as magnetic tape or microfilm. Mr. Napier liked this idea. A five-year inactive period was established as the criterion for removing patient records from the online file.

Next you asked what keys are appropriate for retrieving

patient records. The project team agreed that they want to be able to access patient records by both patient name and patient number.

Chaplain's Report

Mr. Napier requested a report for the Chaplain. Mr. Wilcox had developed a sample report format (see Figure 4-6). Such a report would allow him to more effectively call on and comfort patients, as well as to coordinate bringing in a minister of a patient's own faith if desired. Mr. Napier indicated that the report should be in LOCATION order within RELIGION CODE. A new report heading should be generated for each religion code.

Patient Bill

Mr. Napier discussed the billing document (see Figure 4-7). He explained that the information on the bill was familiar to all, with the following exceptions:

MEMORIAL HOSPITAL
CHAPLAIN'S REPORT
RELIGION CODE 1

LOCATION	PATIENT NAME	AGE
1001–1	PRICE, FRANCIS	25Y
1001–2	PENDER, RONNY	35Y
1002–1	ADAMSON, NANCY	18Y
1003–1	ORR, CHARLES	65Y

Figure 4-6. *Chaplain's Report*

MEMORIAL HOSPITAL
1157 10TH STREET   HOUSTON, TEXAS
713-455-6781

STATEMENT OF ACCOUNT FOR:

GLASS, CHERYL   06-25-80
1218 CLEARVIEW
HOUSTON, TEXAS 77073   ADMIT   05-15-80
                              DISCHARGED   05-16-80
                              PATIENT
                              NUMBER   54321

CODE	ITEM DESCRIPTION	CHARGES	THIRD PARTY COVERAGE	PATIENT SHARE
112	ROOM PRIVATE	80.00	80.00	.00
125	PHARMACY—SOLUTION	25.00	20.00	5.00
112	TELEVISION	4.00	.00	4.00
132	X-RAY—CHEST	16.00	16.00	.00
200	PAYMENT			5.00—
	TOTAL	125.00	116.00	4.00

BALANCE DUE $4.00

Figure 4-7. *Patient Bill*

1. CODE—the same as COST CENTER (see Figure 4-2). The word CODE is used for reference purposes since the cost-center concept is not familiar to most patients. One addition to the previously defined code values is 200, which is used to identify patient payments. Entered payments are subtracted from PATIENT SHARE.
2. CHARGES—the same as AMOUNT (see Figure 4-2).
3. THIRD PARTY COVERAGE—the portion of the charges covered by a third party. This figure is determined and entered by the Accounting Department.
4. PATIENT SHARE—the difference computed by subtracting THIRD PARTY COVERAGE from CHARGES.
5. BALANCE DUE—the total PATIENT SHARE.

Mr. Bunch asked if the billing information should be provided as an additional terminal display. Mr. Napier agreed that it would be helpful to the Accounting Department if an additional display was developed for verifying bills. He pointed out that charges and payments should still be entered using punched cards as discussed previously. You agreed to design a terminal display for the billing information.

This was the final reporting requirement identified by the project team. You suggested that the project team finalize the formats of all reports and that you would present them to the chief administrators of the hospital. The project team agreed. The meeting was concluded.

EXECUTIVE PRESENTATION

When the report formats were finalized, you presented the proposed reporting system at an executive meeting of the chief administrators of the hospital. Mr. Dock and Dr. Cornette were impressed with the proposed system. They identified three additional reports required by management.

First, Mr. Dock said he needs a monthly income analysis report showing dollar volume and transaction volume by cost center. He further indicated the report should identify the dollar volume by the source of payment. After dicussing the report contents, you and Mr. Dock were able to come up with a report format (see Figure 4-8). Mr. Dock explained that MONTHLY TRANSACTIONS are the total number of ITEM DESCRIPTIONS entered by a COST CENTER (see Figure 4-2). DISTRIBUTION is based on the FINANCIAL STATUS code values of the patients receiving services from the COST CENTER.

Dr. Cornette discussed the need for a monthly frequency distribution of FINAL DIAGNOSIS CODE (see "Patient Master Display"). The report need only indicate each possible diagnosis code value and a count of the patients classified in each. You agreed to design a report format.

Mr. Dock and Dr. Cornette indicated they want an exception report enabling them to monitor patient discharges. The report is to show DATE, PATIENT NAME, PATIENT NUMBER, LOCATION, ADMIT DATE, DOCTOR, ADMIT DIAGNOSIS CODE, and AGE. This information is to be printed for a patient under the following conditions. (HINT: This is a good application for a decision table.)

1. If his or her ADMIT DIAGNOSIS CODE (ADC) is between 000000 and 000501, he or she has been in the hospital more

MEMORIAL HOSPITAL
MONTHLY INCOME ANALYSIS

DATE 06-30-80   PAGE 1

COST CENTER	MONTHLY TRANSACTIONS	MONTHLY CHARGES	DISTRIBUTION								
			BLUE CROSS	OTHER-BC	MEDICARE	MEDICAID	COMM INS	WORK COMP	WELFARE	SELF-PAY	OTHER
101	522	18,746.00	10,702.00	2,040.00	568.00	1,120.00	2,561.00	785.00		970.00	
102	68	90,842.75	46,832.25	33,010.50			2,862.00			3,138.00	
103	782	64,781.32	58,651.38	5,480.30						649.64	

Figure 4-8. *Monthly Income Analysis*

94

than three days, and his or her age is less than sixty-five years; *or* if he or she has been in the hospital more than two days and his or her age is sixty-five or greater.

2. If ADC is between 000500 and 001001, hospital stay is more than three days, and age is less than sixty-five; *or* hospital stay is more than five days and age is sixty-five or greater.

3. If ADC is between 001000 and 006001, hospital stay is more than five days, and age is less than sixty-five; *or* hospital stay is more than eight days and age is sixty-five or greater.

4. The patient is currently in intensive care (a message should be printed indicating this).

The report is to be in **PATIENT NUMBER** sequence within **ADC** sequence. You were told to design a report format.

After discussing the last report. Mr. Dock directed you to complete the overall systems specifications for the new system. The meeting was then concluded.

REQUIREMENTS TO COMPLETE CASE

Complete or design all outputs from, and inputs to, the system; create a data element dictionary; construct any necessary decision tables; identify the necessary files; and flowchart the system.

In some instances, the exact format, input media, and validation rules are not told here. You are at liberty to define such issues on a judgmental basis.

There are many ways this case can be expanded into a major time-consuming effort. To avoid this, stay within the requirements of the case.

Chapter 5

Intermountain Distributing Inc.

Distributing companies are engaged primarily in the purchasing of finished goods inventory from manufacturing and/or other distributing companies for resale to retail stores and/or other distributing companies. Accordingly, the primary function of a distributing company is to provide "place" and "time" utility for merchandise by locating and storing merchandise in locations convenient for future sales.

As with all cases in the book, this case is structured to be realistic and representative of the industry for which it is designed. The case is abbreviated so that it can be completed in a short period of time. In particular, the number of reports and terminal displays is reduced. So is the number of data elements. However, the case does include the most important and commonly recognized reporting in the distribution industry.

INTRODUCTION

You have been hired recently as a systems analyst for Intermountain Distributing Inc. The chief executive officers of Intermountain have decided to develop a new information system for the company. You have been assigned as the systems analyst to participate in the project. Mr. Bowman, President of Intermountain, has asked you to attend an executive meeting with the vice-presidents to discuss the new information system.

EXECUTIVE MEETING

At the executive meeting you were presented a copy of Intermountain's organization chart (see Figure 5-1). During the meeting, it was explained to you that Intermountain's operations are based on the following major functions:

- *Sales and Customer Service*—Mr. Wismer, V.P. Marketing, is responsible for this function. It consists of managing sales personnel, advertising, and sales promotion; establishing and executing customer service procedures; and processing sales to the Operations Division.
- *Operations Management*—Mr. Burbback, V.P. Operations, is responsible for this function. It consists of processing orders from the Sales Department; order filling and shipping; and inventory management and purchasing.
- *Financial and Administrative Management*—Ms. Hadley, V.P. Finance and Administration, is responsible for this function. It consists of general ledger, accounts receivable, accounts payable, invoicing, payroll/personnel, and the MIS Department.

Mr. Wismer indicated that it is particularly important to Intermountain to have current, accurate information on the status of orders and inventory. Therefore, the new information system should support online processing of customer orders and the ef-

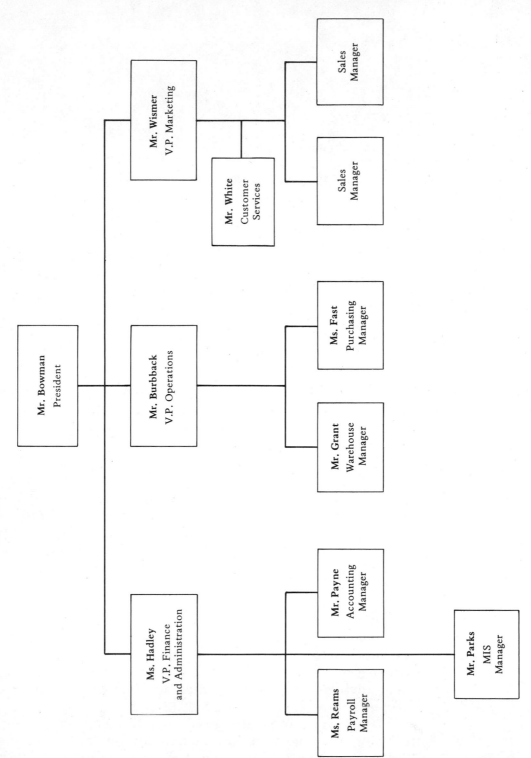

Figure 5-1. *Organization Chart for Intermountain Distributing Inc.*

fects they have on inventory status. Ms. Hadley also expressed her desire that online processing be used to expedite transaction processing and to eliminate the need for a large keypunch staff.

You suggested that a project team be organized to conduct the necessary analysis and design the new information system. You requested that the project team be composed of high-level staff representatives from all major functions affected by the new system.

Mr. Bowman concurred with your suggestion. The following individuals were assigned to the project team:

> Yourself, Systems Analyst and Project Leader
> Mr. Payne, Accounting Manager
> Mr. Grant, Warehouse Manager
> Ms. Fast, Purchasing Manager
> Mr. White, Customer Services

Each team member was charged with determining the information requirements of his or her organizational division. You were assigned overall responsibility for coordinating the team efforts and for developing the systems specifications.

FIRST PROJECT TEAM MEETING

You organized a meeting of the project team to initiate the project. You asked each team member to determine the reporting requirements of his or her division. You explained that once these have been determined, you will be able to "back into" the inputs and processing needed to generate the reports. You also explained that there is cost associated with generating computer-based information and that only information necessary for operations and decision making should be included in the reports.

You indicated that you would coordinate the team efforts to insure integration of the information reporting requirements. For example, inventory status is used by Purchasing for reordering; by the warehouse to determine shipping; by Sales to determine delivery; and by Accounting to value inventory for the general ledger. Accordingly, each department needs to have convenient access to accurate information on inventory status.

SECOND PROJECT TEAM MEETING

After extensive analysis of the information requirements at Inter-mountain, a second project team meeting was called. At this meeting, the project team consolidated the various information requirements into several reports. The reports are presented below.

Inventory Management

Ms. Fast, Purchasing Manager, indicated that the Purchasing and Customer Services Departments require an exception report indicating all inventory items that are out of stock or below their minimum reorder points. Such a report would warn of low stock conditions, facilitate control of stock outages, and allow the Customer Services Department to notify customers so that substitutions or other alternatives can be considered. Ms. Fast provided a sample report as a suggestion for the report format (see Figure 5-2).

You asked Ms. Fast whether VENDOR NUMBER and ITEM

Figure 5-2. Out of Stock/Below Minimum Report

DATE 12–06–80		OUT OF STOCK/ BELOW MINIMUM REPORT					PAGE 1
VENDOR NUMBER	ITEM NUMBER	ITEM DESCRIPTION	STOCK CONDITION	ON HAND	REORDER POINT	QUAN ON ORD	DATE ORDERED
146894	67894220	HAIR DRYER	OUT	****	60	250	12–02–80
792267	68116772	ELECTRIC SHAVER	BELOW MIN	20	40	100	12–03–80
406843	46214558	GUITAR	OUT	****	50	175	12–01–80
787743	22116600	LAMP	BELOW MIN	10	40		12–04–80
432687	98987146	CHESS SET	BELOW MIN	42	100	300	12–01–80

NUMBER are always numeric and whether they are always six and eight digits in length, respectively. Ms. Fast said they are unless an error is made. She also indicated that ITEM DESCRIPTION is up to fifteen characters long, and that REORDER POINT and QUAN ON ORD (quantity on order) are never greater than 999 and 9999, respectively. STOCK CONDITION is to indicate either OUT or BELOW MIN (minimum) depending on the value of ON HAND. Whenever ON HAND equals 0, four asterisks are to be printed below the ON HAND heading. DATE ORDERED is the last date an order was placed—format is MM-DD-YY.

Ms. Fast next indicated that the credit and marketing functions need current information on each inventory item. She proposed that this requirement be accomplished through CRT terminal displays of the entire contents of any inventory record.

All project team members favored this idea. You suggested that this same display could be used for data entry (e.g., additions, deletions, and modifications) to inventory. It was decided, therefore, that online capabilities would be incorporated for update and inquiries of inventory data. You agreed to design the format for the terminal display.

Vendor Reporting

Mr. Grant raised the question of online data entry and retrieval of data about vendors who supply inventory to Intermountain. After determining that vendor data are relatively stable (i.e., do not change often), you suggested that it would be more economical and just as effective to update the data using punched cards. Computer printouts, generated on a periodic basis, could provide the necessary reporting of vendor information. Ms. Fast agreed and further indicated that the only information that Purchasing needs about a vendor is vendor number (a six-digit field), name, address, city, state, zip code, telephone number, and the minimum dollar order the vendor is willing to ship. She indicated that a monthly report containing this information would meet the needs of the Purchasing Department. You agreed to prepare a simple report format for the listing of vendor data.

Invoicing

Mr. Payne, Accounting Manager, and Mr. White, Customer Services, next proposed the contents and format for customer invoices, as illustrated in Figure 5-3. Mr. Payne explained that both a

INTERMOUNTAIN DISTRIBUTING, INC.

DATE 12–06–80
PAGE 01

SOLD TO

THE TOY STORE, INC.
300 BROWN ST.
P.O. BOX 8275
DENVER, COLORADO 76142

SHIP TO

THE TOY STORE
LORETTO PLAZA
415 MAIN ST.
PORTLAND, OREGON 68172

ORDER NO.	123456
INVOICE NO.	123456
CUSTOMER NO.	123456

ITEM NUMBER	DESCRIPTION	U/M	ORDERED	QUANTITY SHIPPED	B/O	RETAIL PRICE	NET PRICE	EXTENDED PRICE
68246789	JET ROCKET	DZ	12	12	0	12.95	6.95	83.40
42687420	BIONIC MAN	EA	10	5	5	8.50	4.40	22.00
68714321	SKATE BOARD	BX	24	0	24	10.25	8.75	.00
77882169	DOLL HOUSE	EA	1	1	0	22.95	11.45	11.45
30002468	GUITAR	EA	6	6	0	165.00	95.50	510.00

TAX	SHIPPING	QUANTITY DISCOUNT	INVOICE TOTAL	RETAIL TOTAL	NET TOTAL
91.33	53.20	275.60	1,918.00	3,848.00	2,049.07

Figure 5-3. *Invoice Format*

billing address (i.e., **SOLD TO**) and a shipping address (i.e., **SHIP TO**) are required. By asking additional questions, you determined the definition of other data elements as follows:

1. DATE—the date the invoice is created.
2. PAGE—the number of this page of the invoice. It is necessary when more than one page is required to list all items ordered.
3. ORDER NO.—a six-digit number assigned to the order when it is entered into the system from a terminal.
4. INVOICE NO.—a six-digit number assigned to the invoice as it is printed. For each batch of invoices on a given day, the

invoice number is initialized to 1, and incremented by 1 for each invoice generated. Thus, given a date and invoice number, a specific invoice can be identified.

5. CUSTOMER NO.—a six-digit number assigned to each customer by the Accounting Department.

6. U/M—the unit of measure for an item. Options are:

EA = Each
BX = Box
DZ = Dozen
FT = Foot
YD = Yard

6. QUANTITY ORDERED, SHIPPED, and B/O (Back Ordered)— three four-digit fields. Back orders apply to items not currently in stock that have to be shipped at a later date.

7. RETAIL PRICE—the suggested retail price for each item. It should never exceed $2,000.00.

8. NET PRICE—the price to the buyer. It should never exceed $2000.00.

9. EXTENDED PRICE—the price for the total number of units shipped (i.e., QUANTITY SHIPPED × NET PRICE). It should never exceed $10,000.00.

10. NET TOTAL—the total of the EXTENDED PRICE column.

11. RETAIL TOTAL—the total of RETAIL PRICE × QUANTITY SHIPPED for each item.

12. QUANTITY DISCOUNT—the total quantity discount for the order. It is computed as follows. Each inventory item has up to four quantity break levels. The levels and their discounts follow:

Quantity Level A = 1.0%
Quantity Level B = 1.5%
Quantity Level C = 2.0%
Quantity Level D = 3.0%

The specific numeric value assigned to a quantity level is determined by the V.P. of Marketing and can vary among inventory items. For example, quantity break levels may be assigned to two inventory items as follows:

Quantity Level	Guitars	Guitar Picks
A	5	100
B	10	250
C	15	500
D	20	1000

During invoice preparation, the computer must compare the quantity ordered with the quantity break level values to determine the percent to use for computing the quantity discount for each item ordered. The total discount for all items on an invoice determines the QUANTITY DISCOUNT for the invoice. (HINT: This is a good application for a decision table.)

13. SHIPPING = 3% of (NET TOTAL − QUANTITY DISCOUNT).
14. TAX = 5% of (NET TOTAL − QUANTITY DISCOUNT + SHIPPING).
15. INVOICE TOTAL = NET TOTAL − QUANTITY DISCOUNT + SHIPPING + TAX.

You asked if quantity-level data (see QUANTITY DISCOUNT) should be added to the CRT display for inventory items. Ms. Fast said that was an excellent idea and to please do so.

Mr. Payne indicated that Ms. Hadley had requested a daily customer transaction report to monitor sales activity. The report should contain one line for each invoice and report the invoice number, customer number, customer name, retail total, net total, quantity discount, tax, shipping charge, and invoice total. An appropriate heading should be provided. Mr. Payne asked you to design the report format.

With Mr. Payne's request, the project team meeting was concluded. A third project team meeting was scheduled to discuss warehouse and credit requirements.

THIRD PROJECT TEAM MEETING

Filling and Shipping Orders

At the third project team meeting, Mr. Grant discussed the requirements for filling and shipping orders. He indicated that a picking list and a packing slip are required for processing orders in the warehouse. Mr. Grant explained the picking list is used to determine the items ordered by a customer and the bin locations in the warehouse where the items are stored. He proposed the format illustrated in Figure 5-4 for the picking list.

You asked Mr. Grant to explain the format of the BIN LOCATION. Mr. Grant explained that the first character indicated the warehouse and it could be coded as an A, B, or C. The warehouse code is followed by a blank and four digits separated by a dash. The

DATE 12–08–80		PICKING LIST			PAGE 01
CUSTOMER NO. 123456		**CUSTOMER NAME** THE HANDY HUT		**ORDER NO.** 123456	

BIN LOCATION	ITEM NUMBER	ITEM DESCRIPTION	U/M	SHIP	WEIGHT
A 01–13	46872431	LIPSTICK	BX	5	1.2
A 01–17	61127421	HAIR SPRAY	BX	10	4.0
A 03–04	77864230	HAIR DRYER	EA	20	24.6
B 07–06	46789106	SIDE MOLDING	FT	100	.5
•	•	•	•	•	•
•	•		•	•	•
	•		•		

TOTAL WEIGHT 121.6

Figure 5-4.   *Picking List*

first two digits indicate the aisle number which ranges from 01 to 50. The last two digits indicate the bin numbers within an aisle. Bin numbers range from 01 to 99.

The packing slip is enclosed in the package for shipping. It is used by the customer to validate the goods received. The format that Mr. Grant proposed for the packing slip is illustrated in Figure 5-5. Mr. Grant explained that WEIGHT is the weight of the U/M (unit of measure).

After analyzing the picking list and the packing slip, you noticed they were quite similar in content. You raised the question of consolidating the contents in one document. This could be accomplished by adding the bin location to the packing slip.

Mr. Grant asked what advantages this approach offered. You explained that by consolidating the contents, the two documents could be printed at the same time (i.e., by using multiple-part paper). This would save computer time and expedite the processing of picking and packing information needed at the warehouse. Mr. Grant and the other project team members accepted your sugges-

DATE 12–08–80		PACKING SLIP			PAGE 01	

CUSTOMER NO. 123456

SOLD TO	SHIP TO	ORDER NO.	123456
		ORDER DATE	12–01–80
THE HANDY HUT	THE HANDY HUT		
CENTER BUILDING	CENTER BUILDING		
P.O. BOX 8172	P.O. BOX 8172		
ATLANTA, GEORGIA 71682	ATLANTA, GEORGIA 71682		

ITEM NUMBER	ITEM DESCRIPTION	U/M	SHIPPED	WEIGHT	SUGGESTED RETAIL
46872431	LIPSTICK	BX	5	1.2	1.29
61127421	HAIR SPRAY	BX	10	4.0	1.95
77864230	HAIR DRYER	EA	20	24.6	12.75
46789106	SIDE MOLDING	FT	100	.5	8.95

TOTAL WEIGHT 121.6

Figure 5-5. *Packing Slip*

tion. You agreed to design a new combined picking list/packing slip document.

Customer and Credit Reporting

The next topic discussed at the meeting was the customer information needed by the accounting and marketing functions. Mr. Payne explained that some of the required customer data had been identified in previously defined reporting (e.g., customer number, customer name, billing and shipping addresses, and order data). However, the Accounts Receivable Department has to approve credit for each order processed. Credit approval consists of adding the dollar amount of each incoming order to the customer's current balance. This new balance is then compared to the customer's credit limit. If the new balance is less than the credit limit, the order is approved. Otherwise, the determination of approval or rejection of credit is handled on an exception basis. Each exception usually involves the supervisor of Accounts Receivable, who in turn contacts Mr. White in Customer Services, and possibly the customer, to decide an appropriate course of action.

You asked Mr. Payne how customer payments are handled. He explained that customer payments are cycled through the Accounts Receivable Department and deducted from customer balances. He said it would be nice if the new system allowed accounts receivable personnel to deduct payments from balances via computer terminals. You agreed to this approach.

Next, you asked Mr. Payne if it would help the efficiency of the Accounts Receivable Department if the computer automatically processed routine credit authorizations. All exceptions could be reported, via hard-copy terminals, to the Accounts Receivable Department for special disposition.

Mr. Payne was agreeable to this plan. He asked how such a system would operate. You explained that all orders would be entered into the system by order processing clerks using CRT terminals in their offices. As the orders were entered, the computer system would perform the credit checks. If credit was approved, the order would be admitted to the system. If credit was not approved, the system would notify the order processing clerk to hold the particular order until further notification from the Accounts Receivable Department. The system would then generate an exception report on a hard-copy terminal in the Accounts Receivable Department. The report would be used to resolve the credit problem. Once a decision had been made, the Accounts Receivable Department could call the Order Processing Department and instruct them to re-enter the order or to deliver it to Customer Services at the end of the day depending on whether or not the order is approved.

Mr. Payne was impressed with this approach. You asked him to define the information required on the exception report. He listed the following items:

1. CUSTOMER NAME
2. CUSTOMER NUMBER
3. CUSTOMER PHONE (area code included)
4. YEAR ACCOUNT OPENED—A six-digit field of the form MM-DD-YY.
5. CREDIT LIMIT—not to exceed $450,000.00.
6. CREDIT RATING—A one digit field coded as follows: 1 = Excellent, 2 = Good, 3 = Poor.

7. CREDIT RATING DATE—A six-digit field of the form MM-DD-YY.
8. DATE OF LAST SALE—MM-DD-YY.
9. CURRENT BALANCE—not to exceed $999,999.00.
10. CURRENT ORDER AMOUNT—the amount of the order that, when added to the current balance, caused the exception report to occur.

You agreed to develop a simple format suitable for a hard-copy terminal to report credit authorization exceptions to the Accounts Receivable Department.

This was the final reporting requirement identified by the project team. You suggested that the project team finalize the formats of all reports and that you would present them to the executives of Intermountain. The project team members agreed. The meeting was concluded.

EXECUTIVE PRESENTATION

When the report formats were finalized, you presented the proposed reporting system at an executive meeting. Mr. Bowman and the vice-presidents were impressed with the proposed system. They identified two additional reports required by management.

First, Mr. Bowman and Mr. Burbback discussed the need for a summary report indicating the profitability of each inventory item. They said the report should indicate past-month and year-to-date totals for each inventory item. The totals to be reported follow:

1. UNITS SOLD—total number of units sold.
2. UNIT OF MEASURE—measure used to define unit (e.g., box, dozen, each).
3. NUMBER OF INVOICES—total number of invoices prepared.
4. GROSS SALES—total dollar value of items sold (i.e., units sold × unit price).
5. DISCOUNTS—total of all discounts as computed during invoice generation.
6. SALES EXPENSE—computed as follows:
    NUMBER OF INVOICES × $36.00 + 15% of GROSS SALES.
7. NET PROFIT—GROSS SALES less SALES EXPENSE and DISCOUNTS.
8. % ITEM PROFIT—NET PROFIT divided by GROSS SALES.

Mr. Burbback explained that the profitability-by-item report should be generated at the end of each month. Year-to-date totals should be initialized to zero at the end of the calendar year. Past-month totals should be computed each month and added to the year-to-date totals.

Mr. Bowman asked you to design an appropriate format for the report. You agreed to do so.

Next, Mr. Wismer and Mr. Bowman presented the need for a profitability-by-customer report. They indicated that the report should provide year-to-date and past-month totals for each customer as follows:

1. NUMBER OF INVOICES
2. GROSS SALES
3. DISCOUNTS
4. SALES EXPENSE
5. NET PROFIT
6. % CUSTOMER PROFIT—NET PROFIT divided by GROSS SALES.

The initialization and monthly adjustments to totals are identical to those for the profitability-by-item report. You agreed to design a report format.

After discussing the two management reports, Mr. Bowman directed you to complete the overall systems specifications for the new system. The meeting was then concluded.

REQUIREMENTS TO COMPLETE CASE

Complete or design all outputs from, and inputs to, the system; create a data element dictionary; construct any necessary decision tables; identify the necessary files; and flowchart the system.

In some instances, the exact format and validation rules are not specified here. You are at liberty to define such issues on a judgmental basis.

There are many ways this case can be expanded into a major time-consuming effort. To avoid this, stay within the requirements defined in the case.

REPORT DEFINITION for
_____ Application

I REPORT LAYOUT

NAME OF REPORT _____
PREPARED BY _____
DATE _____ PAGE _____ OF _____

REPORT DEFINITION for
_____ Application

NAME OF REPORT _____
PREPARED BY _____
DATE _____ PAGE _____ OF _____

I REPORT LAYOUT

I REPORT LAYOUT

REPORT DEFINITION for
_____ **Application**

I REPORT LAYOUT

NAME OF REPORT _____

PREPARED BY _____
DATE _____ PAGE _____ OF _____

I REPORT LAYOUT

NAME OF REPORT _____

PREPARED BY _____

DATE _____ PAGE ____ OF ____

REPORT DEFINITION for

_____ **Application**

I REPORT LAYOUT

NAME OF REPORT _____

PREPARED BY _____

DATE _____ PAGE _____ OF _____

REPORT DEFINITION for
_____ Application

I REPORT LAYOUT

NAME OF REPORT _____
PREPARED BY _____
DATE _____ PAGE ____ OF ____

REPORT DEFINITION for
_____
_____ Application

I REPORT LAYOUT

REPORT DEFINITION
NAME OF REPORT _____
PREPARED BY _____
DATE _____ PAGE _____ OF _____

REPORT DEFINITION for
_____ Application

I REPORT LAYOUT

REPORT DEFINITION for

NAME OF REPORT
PREPARED BY
DATE _____ PAGE ___ OF ___

I REPORT LAYOUT

REPORT DEFINITION for
_____ Application

NAME OF REPORT _____
PREPARED BY _____
DATE _____ PAGE _____ OF _____

I REPORT LAYOUT

NAME OF REPORT _____

PREPARED BY _____

DATE _____   PAGE _____ OF _____

REPORT DEFINITION for _____ Application

I REPORT LAYOUT

NAME OF REPORT _____
PREPARED BY _____
DATE _____ PAGE _____ Of _____

REPORT DEFINITION for
_____ Application

I REPORT LAYOUT

NAME OF REPORT _____
PREPARED BY _____
DATE _____ PAGE ___ OF ___

REPORT DEFINITION for

_____ Application

I REPORT LAYOUT

REPORT DEFINITION for

NAME OF REPORT _____

PREPARED BY _____

DATE _____ PAGE _____ OF _____

REPORT DEFINITION for

_____ Application

I REPORT LAYOUT

REPORT LAYOUT

NAME OF REPORT _____

PREPARED BY _____

DATE _____ PAGE _____ OF _____

REPORT DEFINITION for
_____ Application

I REPORT LAYOUT

NAME OF REPORT _____
PREPARED BY _____
DATE _____ PAGE ____ OF ____

REPORT DEFINITION for

_____ Application

I REPORT LAYOUT

REPORT DEFINITION

NAME OF REPORT _____

PREPARED BY _____

DATE _____ PAGE _____ OF _____

REPORT DEFINITION for
_____
_____ Application

I REPORT LAYOUT

NAME OF REPORT _____
PREPARED BY _____
DATE _____ PAGE _____ OF _____

I REPORT LAYOUT

NAME OF REPORT _____

PREPARED BY _____

DATE _____ PAGE _____ OF _____

REPORT DEFINITION for
_____ Application

I REPORT LAYOUT

NAME OF REPORT _____
PREPARED BY _____
DATE _____ PAGE ___ OF ___

I REPORT LAYOUT

NAME OF REPORT _____

PREPARED BY _____

DATE _____ PAGE ____ OF ____

NAME OF FORMAT: _____

| 1 2 3 4 5 6 7 8 9 10 11 12 13 14 15 16 17 18 19 20 21 22 23 24 25 26 27 28 29 30 31 32 33 34 35 36 37 38 39 40 41 42 43 44 45 46 47 48 49 50 |

NAME OF FORMAT: _____

| 1 2 3 4 5 6 7 8 9 10 11 12 13 14 15 16 17 18 19 20 21 22 23 24 25 26 27 28 29 30 31 32 33 34 35 36 37 38 39 40 41 42 43 44 45 46 47 48 49 50 |

NAME OF FORMAT: _____

| 1 2 3 4 5 6 7 8 9 10 11 12 13 14 15 16 17 18 19 20 21 22 23 24 25 26 27 28 29 30 31 32 33 34 35 36 37 38 39 40 41 42 43 44 45 46 47 48 49 50 |

| 1 2 3 4 5 6 7 8 9 10 11 12 13 14 15 16 17 18 19 20 21 22 23 24 25 26 27 28 29 30 31 32 33 34 35 36 37 38 39 40 41 42 43 44 45 46 47 48 49 50 |

I REPORT LAYOUT

REPORT DEFINITION for _____ Application

NAME OF REPORT: _____
PREPARED BY _____
DATE _____ PAGE _____ OF _____

1 2 3 4 5 6 7 8 9 10 11 12 13 14 15 16 17 18 19 20 21 22 23 24 25 26 27 28 29 30 31 32 33 34 35 36 37 38 39 40 41 42 43 44 45 46 47 48 49 50

NAME OF FORMAT: _____

1 2 3 4 5 6 7 8 9 10 11 12 13 14 15 16 17 18 19 20 21 22 23 24 25 26 27 28 29 30 31 32 33 34 35 36 37 38 39 40 41 42 43 44 45 46 47 48 49 50

NAME OF FORMAT: _____

1 2 3 4 5 6 7 8 9 10 11 12 13 14 15 16 17 18 19 20 21 22 23 24 25 26 27 28 29 30 31 32 33 34 35 36 37 38 39 40 41 42 43 44 45 46 47 48 49 50

NAME OF FORMAT: _____

1 2 3 4 5 6 7 8 9 10 11 12 13 14 15 16 17 18 19 20 21 22 23 24 25 26 27 28 29 30 31 32 33 34 35 36 37 38 39 40 41 42 43 44 45 46 47 48 49 50

NAME OF FORMAT: _____

NAME OF FORMAT:

1 2 3 4 5 6 7 8 9 10 11 12 13 14 15 16 17 18 19 20 21 22 23 24 25 26 27 28 29 30 31 32 33 34 35 36 37 38 39 40 41 42 43 44 45 46 47 48 49 50

NAME OF FORMAT:

1 2 3 4 5 6 7 8 9 10 11 12 13 14 15 16 17 18 19 20 21 22 23 24 25 26 27 28 29 30 31 32 33 34 35 36 37 38 39 40 41 42 43 44 45 46 47 48 49 50

NAME OF FORMAT:

1 2 3 4 5 6 7 8 9 10 11 12 13 14 15 16 17 18 19 20 21 22 23 24 25 26 27 28 29 30 31 32 33 34 35 36 37 38 39 40 41 42 43 44 45 46 47 48 49 50

1 2 3 4 5 6 7 8 9 10 11 12 13 14 15 16 17 18 19 20 21 22 23 24 25 26 27 28 29 30 31 32 33 34 35 36 37 38 39 40 41 42 43 44 45 46 47 48 49 50

I REPORT LAYOUT

NAME OF REPORT _____
PREPARED BY _____
DATE _____
PAGE ___ OF ___

NAME OF FORMAT:

NAME OF FORMAT:

NAME OF FORMAT:

I REPORT LAYOUT

NAME OF REPORT _____

PREPARED BY _____

DATE _____ PAGE ____ OF ____

NAME OF FORMAT:

```
1 2 3 4 5 6 7 8 9 10 11 12 13 14 15 16 17 18 19 20 21 22 23 24 25 26 27 28 29 30 31 32 33 34 35 36 37 38 39 40 41 42 43 44 45 46 47 48 49 50
```

NAME OF FORMAT:

```
1 2 3 4 5 6 7 8 9 10 11 12 13 14 15 16 17 18 19 20 21 22 23 24 25 26 27 28 29 30 31 32 33 34 35 36 37 38 39 40 41 42 43 44 45 46 47 48 49 50
```

NAME OF FORMAT:

```
1 2 3 4 5 6 7 8 9 10 11 12 13 14 15 16 17 18 19 20 21 22 23 24 25 26 27 28 29 30 31 32 33 34 35 36 37 38 39 40 41 42 43 44 45 46 47 48 49 50
```

NAME OF FORMAT:

```
1 2 3 4 5 6 7 8 9 10 11 12 13 14 15 16 17 18 19 20 21 22 23 24 25 26 27 28 29 30 31 32 33 34 35 36 37 38 39 40 41 42 43 44 45 46 47 48 49 50
```

I REPORT LAYOUT

REPORT DEFINITION _____
PAGE ____ OF ____

NAME OF REPORT _____
PREPARED BY _____
DATE _____

1 2 3 4 5 6 7 8 9 10 11 12 13 14 15 16 17 18 19 20 21 22 23 24 25 26 27 28 29 30 31 32 33 34 35 36 37 38 39 40 41 42 43 44 45 46 47 48 49 50

NAME OF FORMAT:

1 2 3 4 5 6 7 8 9 10 11 12 13 14 15 16 17 18 19 20 21 22 23 24 25 26 27 28 29 30 31 32 33 34 35 36 37 38 39 40 41 42 43 44 45 46 47 48 49 50

NAME OF FORMAT:

1 2 3 4 5 6 7 8 9 10 11 12 13 14 15 16 17 18 19 20 21 22 23 24 25 26 27 28 29 30 31 32 33 34 35 36 37 38 39 40 41 42 43 44 45 46 47 48 49 50

NAME OF FORMAT:

1 2 3 4 5 6 7 8 9 10 11 12 13 14 15 16 17 18 19 20 21 22 23 24 25 26 27 28 29 30 31 32 33 34 35 36 37 38 39 40 41 42 43 44 45 46 47 48 49 50

NAME OF FORMAT:

REPORT DEFINITION for
_____
_____
Application

I REPORT LAYOUT

NAME OF REPORT _____
PREPARED BY _____
DATE _____ PAGE ____ OF ____

1 2 3 4 5 6 7 8 9 10 11 12 13 14 15 16 17 18 19 20 21 22 23 24 25 26 27 28 29 30 31 32 33 34 35 36 37 38 39 40 41 42 43 44 45 46 47 48 49 50

NAME OF FORMAT:

1 2 3 4 5 6 7 8 9 10 11 12 13 14 15 16 17 18 19 20 21 22 23 24 25 26 27 28 29 30 31 32 33 34 35 36 37 38 39 40 41 42 43 44 45 46 47 48 49 50

NAME OF FORMAT:

1 2 3 4 5 6 7 8 9 10 11 12 13 14 15 16 17 18 19 20 21 22 23 24 25 26 27 28 29 30 31 32 33 34 35 36 37 38 39 40 41 42 43 44 45 46 47 48 49 50

NAME OF FORMAT:

1 2 3 4 5 6 7 8 9 10 11 12 13 14 15 16 17 18 19 20 21 22 23 24 25 26 27 28 29 30 31 32 33 34 35 36 37 38 39 40 41 42 43 44 45 46 47 48 49 50

NAME OF FORMAT:

I REPORT LAYOUT

NAME OF REPORT _____
_____
PREPARED BY _____
DATE _____
PAGE ___ OF ___

1 2 3 4 5 6 7 8 9 10 11 12 13 14 15 16 17 18 19 20 21 22 23 24 25 26 27 28 29 30 31 32 33 34 35 36 37 38 39 40 41 42 43 44 45 46 47 48 49 50

NAME OF FORMAT:

1 2 3 4 5 6 7 8 9 10 11 12 13 14 15 16 17 18 19 20 21 22 23 24 25 26 27 28 29 30 31 32 33 34 35 36 37 38 39 40 41 42 43 44 45 46 47 48 49 50

NAME OF FORMAT:

1 2 3 4 5 6 7 8 9 10 11 12 13 14 15 16 17 18 19 20 21 22 23 24 25 26 27 28 29 30 31 32 33 34 35 36 37 38 39 40 41 42 43 44 45 46 47 48 49 50

NAME OF FORMAT:

1 2 3 4 5 6 7 8 9 10 11 12 13 14 15 16 17 18 19 20 21 22 23 24 25 26 27 28 29 30 31 32 33 34 35 36 37 38 39 40 41 42 43 44 45 46 47 48 49 50

I REPORT LAYOUT

NAME OF REPORT _____

PREPARED BY _____

DATE _____ PAGE ___ OF ___

| 1 2 3 4 5 6 7 8 9 10 11 12 13 14 15 16 17 18 19 20 21 22 23 24 25 26 27 28 29 30 31 32 33 34 35 36 37 38 39 40 41 42 43 44 45 46 47 48 49 50 |

NAME OF FORMAT: _____

| 1 2 3 4 5 6 7 8 9 10 11 12 13 14 15 16 17 18 19 20 21 22 23 24 25 26 27 28 29 30 31 32 33 34 35 36 37 38 39 40 41 42 43 44 45 46 47 48 49 50 |

NAME OF FORMAT: _____

| 1 2 3 4 5 6 7 8 9 10 11 12 13 14 15 16 17 18 19 20 21 22 23 24 25 26 27 28 29 30 31 32 33 34 35 36 37 38 39 40 41 42 43 44 45 46 47 48 49 50 |

NAME OF FORMAT: _____

| 1 2 3 4 5 6 7 8 9 10 11 12 13 14 15 16 17 18 19 20 21 22 23 24 25 26 27 28 29 30 31 32 33 34 35 36 37 38 39 40 41 42 43 44 45 46 47 48 49 50 |

I REPORT LAYOUT

NAME OF REPORT _____
PREPARED BY _____
DATE _____ PAGE _____ OF _____

1 2 3 4 5 6 7 8 9 10 11 12 13 14 15 16 17 18 19 20 21 22 23 24 25 26 27 28 29 30 31 32 33 34 35 36 37 38 39 40 41 42 43 44 45 46 47 48 49 50

NAME OF FORMAT:

1 2 3 4 5 6 7 8 9 10 11 12 13 14 15 16 17 18 19 20 21 22 23 24 25 26 27 28 29 30 31 32 33 34 35 36 37 38 39 40 41 42 43 44 45 46 47 48 49 50

NAME OF FORMAT:

1 2 3 4 5 6 7 8 9 10 11 12 13 14 15 16 17 18 19 20 21 22 23 24 25 26 27 28 29 30 31 32 33 34 35 36 37 38 39 40 41 42 43 44 45 46 47 48 49 50

NAME OF FORMAT:

1 2 3 4 5 6 7 8 9 10 11 12 13 14 15 16 17 18 19 20 21 22 23 24 25 26 27 28 29 30 31 32 33 34 35 36 37 38 39 40 41 42 43 44 45 46 47 48 49 50

I REPORT LAYOUT

REPORT DEFINITION

NAME OF REPORT _____

PREPARED BY _____

DATE _____   PAGE ____ OF ____

1 2 3 4 5 6 7 8 9 10 11 12 13 14 15 16 17 18 19 20 21 22 23 24 25 26 27 28 29 30 31 32 33 34 35 36 37 38 39 40 41 42 43 44 45 46 47 48 49 50

NAME OF FORMAT:

1 2 3 4 5 6 7 8 9 10 11 12 13 14 15 16 17 18 19 20 21 22 23 24 25 26 27 28 29 30 31 32 33 34 35 36 37 38 39 40 41 42 43 44 45 46 47 48 49 50

NAME OF FORMAT:

1 2 3 4 5 6 7 8 9 10 11 12 13 14 15 16 17 18 19 20 21 22 23 24 25 26 27 28 29 30 31 32 33 34 35 36 37 38 39 40 41 42 43 44 45 46 47 48 49 50

NAME OF FORMAT:

1 2 3 4 5 6 7 8 9 10 11 12 13 14 15 16 17 18 19 20 21 22 23 24 25 26 27 28 29 30 31 32 33 34 35 36 37 38 39 40 41 42 43 44 45 46 47 48 49 50

REPORT DEFINITION for

_____

_____ Application

I REPORT LAYOUT

| | | | | | | | | | | | | | | | | | | | | | | | | | | | | | | | | | | | | | | | | | | | | | | | |
1 2 3 4 5 6 7 8 9 10 11 12 13 14 15 16 17 18 19 20 21 22 23 24 25 26 27 28 29 30 31 32 33 34 35 36 37 38 39 40 41 42 43 44 45 46 47 48 49 50

NAME OF FORMAT: _____

| | | | | | | | | | | | | | | | | | | | | | | | | | | | | | | | | | | | | | | | | | | | | | | | |
1 2 3 4 5 6 7 8 9 10 11 12 13 14 15 16 17 18 19 20 21 22 23 24 25 26 27 28 29 30 31 32 33 34 35 36 37 38 39 40 41 42 43 44 45 46 47 48 49 50

NAME OF FORMAT: _____

| | | | | | | | | | | | | | | | | | | | | | | | | | | | | | | | | | | | | | | | | | | | | | | | |
1 2 3 4 5 6 7 8 9 10 11 12 13 14 15 16 17 18 19 20 21 22 23 24 25 26 27 28 29 30 31 32 33 34 35 36 37 38 39 40 41 42 43 44 45 46 47 48 49 50

NAME OF FORMAT: _____

| | | | | | | | | | | | | | | | | | | | | | | | | | | | | | | | | | | | | | | | | | | | | | | | |
1 2 3 4 5 6 7 8 9 10 11 12 13 14 15 16 17 18 19 20 21 22 23 24 25 26 27 28 29 30 31 32 33 34 35 36 37 38 39 40 41 42 43 44 45 46 47 48 49 50

DECISION TABLE FOR _____ APPLICATION

DECISION TABLE NAME _____ REFERENCE NO. _____

PREPARED BY _____ DATE _____

DECISION RULES

CONDITIONS/
COURSES OF
ACTION

01	02	03	04	05	06	07	08	09	10	11	12	13	14	15	16	17	18	19	20	21	22	23	24	25	26	27	28	29	30	31	32	33	34	35	36	37	38	39	40	41	42	43	44	45

DECISION TABLE FOR _____ APPLICATION

DECISION TABLE NAME _____ REFERENCE NO. _____

PREPARED BY _____ DATE _____

DECISION RULES

CONDITIONS/ COURSES OF ACTION	01	02	03	04	05	06	07	08	09	10	11	12	13	14	15	16	17	18	19	20	21	22	23	24	25	26	27	28	29	30	31	32	33	34	35	36	37	38	39	40	41	42	43	44	45

DECISION TABLE FOR _____ APPLICATION

DECISION TABLE NAME _____ REFERENCE NO. _____

PREPARED BY _____ DATE _____

DECISION RULES

CONDITIONS/ COURSES OF ACTION

| 01 | 02 | 03 | 04 | 05 | 06 | 07 | 08 | 09 | 10 | 11 | 12 | 13 | 14 | 15 | 16 | 17 | 18 | 19 | 20 | 21 | 22 | 23 | 24 | 25 | 26 | 27 | 28 | 29 | 30 | 31 | 32 | 33 | 34 | 35 | 36 | 37 | 38 | 39 | 40 | 41 | 42 | 43 | 44 | 45 |

DECISION TABLE FOR _____ APPLICATION

DECISION TABLE NAME _____ REFERENCE NO. _____

PREPARED BY _____ DATE _____

DECISION RULES

CONDITIONS/ COURSES OF ACTION	01	02	03	04	05	06	07	08	09	10	11	12	13	14	15	16	17	18	19	20	21	22	23	24	25	26	27	28	29	30	31	32	33	34	35	36	37	38	39	40	41	42	43	44	45

DECISION TABLE FOR _____ APPLICATION

DECISION TABLE NAME _____ REFERENCE NO. _____

PREPARED BY _____ DATE _____

CONDITIONS/
COURSES OF
ACTION

DECISION RULES

01	02	03	04	05	06	07	08	09	10	11	12	13	14	15	16	17	18	19	20	21	22	23	24	25	26	27	28	29	30	31	32	33	34	35	36	37	38	39	40	41	42	43	44	45

DECISION TABLE FOR _____ APPLICATION _____

DECISION TABLE NAME _____ REFERENCE NO. _____

PREPARED BY _____ DATE _____

DECISION RULES

CONDITIONS/
COURSES OF
ACTION

01	02	03	04	05	06	07	08	09	10	11	12	13	14	15	16	17	18	19	20	21	22	23	24	25	26	27	28	29	30	31	32	33	34	35	36	37	38	39	40	41	42	43	44	45

DECISION TABLE FOR _____ APPLICATION _____

DECISION TABLE NAME _____ REFERENCE NO. _____

PREPARED BY _____ DATE _____

CONDITIONS/
COURSES OF
ACTION

DECISION RULES

01	02	03	04	05	06	07	08	09	10	11	12	13	14	15	16	17	18	19	20	21	22	23	24	25	26	27	28	29	30	31	32	33	34	35	36	37	38	39	40	41	42	43	44	45

DECISION TABLE FOR _____ APPLICATION

DECISION TABLE NAME _____ REFERENCE NO._____

PREPARED BY _____ DATE_____

CONDITIONS/
COURSES OF
ACTION

DECISION RULES

01	02	03	04	05	06	07	08	09	10	11	12	13	14	15	16	17	18	19	20	21	22	23	24	25	26	27	28	29	30	31	32	33	34	35	36	37	38	39	40	41	42	43	44	45

DECISION RULES

01	02	03	04	05	06	07	08	09	10	11	12	13	14	15	16	17	18	19	20	21	22	23	24	25	26	27	28	29	30	31	32	33	34	35	36	37	38	39	40	41	42	43	44	45

CONDITIONS/
COURSES OF
ACTION

DECISION TABLE FOR _____ APPLICATION

DECISION TABLE NAME _____ REFERENCE NO. _____

PREPARED BY _____ DATE _____

CONDITIONS/
COURSES OF
ACTION

DECISION RULES

01	02	03	04	05	06	07	08	09	10	11	12	13	14	15	16	17	18	19	20	21	22	23	24	25	26	27	28	29	30	31	32	33	34	35	36	37	38	39	40	41	42	43	44	45

DECISION TABLE FOR _____ APPLICATION

DECISION TABLE NAME _____ REFERENCE NO. _____

PREPARED BY _____ DATE _____

CONDITIONS/ COURSES OF ACTION	DECISION RULES																																											
	01	02	03	04	05	06	07	08	09	10	11	12	13	14	15	16	17	18	19	20	21	22	23	24	25	26	27	28	29	30	31	32	33	34	35	36	37	38	39	40	41	42	43	44	45

DECISION TABLE FOR _____ APPLICATION

DECISION TABLE NAME _____ REFERENCE NO. _____

PREPARED BY _____ DATE _____

DECISION RULES

	01	02	03	04	05	06	07	08	09	10	11	12	13	14	15	16	17	18	19	20	21	22	23	24	25	26	27	28	29	30	31	32	33	34	35	36	37	38	39	40	41	42	43	44	45

CONDITIONS/
COURSES OF
ACTION